Graph Theory and Applications

Graph Theory and Applications

with Exercises and Problems

Jean-Claude Fournier

ISTE

WILEY

First published in France in 2006 by Hermes Science/Lavoisier entitled *Théorie des graphes et applications, avec exercices et problèmes* © LAVOISIER, 2006
First published in Great Britain and the United States in 2009 by ISTE Ltd and John Wiley & Sons, Inc.

ISTE Ltd
27-37 St George's Road
London SW19 4EU
UK

www.iste.co.uk

John Wiley & Sons, Inc.
111 River Street
Hoboken, NJ 07030
USA

www.wiley.com

Library of Congress Cataloging-in-Publication Data

Fournier, Jean-Claude.
[Théorie des graphes et applications, avec exercices et problèmes. English]
Graph theory and applications with exercises and problems / Jean-Claude Fournier.
p. cm.
Includes bibliographical references and index.
ISBN 978-1-84821-070-7
1. Graph theory. 2. Graph theory--Problems, exercises, etc. I. Title.
QA166.F68513 2009
511'.5--dc22

2008043204

British Library Cataloguing-in-Publication Data
A CIP record for this book is available from the British Library
ISBN: 978-1-84821-070-7

To Hugo, Eliott, Mathieu,
Elise, Aurélie, Antonin, Ethan
and those to come ...

Table of Contents

Introduction

The concept of a graph is relatively recent since it only formally appeared during the 20th century. Today it has become essential in many fields, in particular in applied and fundamental computer science, in optimization, and in algorithmic complexity. The study of graphs and their applications therefore provides an opportunity to deal with very diverse questions with numerous applications. It can be used, for example, to develop task scheduling methods from optimal paths in graphs, or properties of communication networks in relation to the connectedness of graphs. Historically, graphs were considered long before the theory was developed, with famous problems such as the Königsberg bridge problem (presented in Chapter 9).

This work studies the principal aspects of graph theory with special emphasis on major applications. Its content is aimed at advanced undergraduate and students beginning post-graduate studies in mathematics and computer science. It calls for very few pre-requisites, essentially a familiarity with basic vocabulary and mathematical reasoning. In return, the study of this new subject will be a great opportunity for students to test and improve their personal logic. Graph theory is indeed a new subject, which is very different from the traditional mathematics taught, but which requires the same intellectual rigor. The great novelty of this subject may throw even the best students in mathematics, which shows how beneficial studying this subject can be!

This book, conceived as a textbook, is the result of long experience teaching bachelor and masters level students (notably as part of the French master in applied information technology: "MIAGE"). It also contains many exercises and problems classified in five levels of increasing difficulty:

1. Certain points and quite easy proofs are left to the reader's initiative. They are indicated in the margin by a †.

2. Among the exercises given at the end of each chapter, some are marked with a +, indicating a useful complement to the chapter, which may also be used elsewhere in the text.

3. The exercises which are not marked are of normal difficulty. They are standard exercises, among which you will find the most classic in the field.

4. The exercises marked with a * are more difficult. They go into a subject in more depth than the chapter does.

5. A last level with a few problems is given at the end, with commentaries and sometimes hints for their resolution (Chapter 12).

Here is some information on how this book is organized. Except for the two chapters on definitions and basic concepts, which are Chapter 1 for undirected graphs and Chapter 4 for directed graphs, each chapter presents a specific subject with a major application, and is therefore relatively independent from the other chapters. It is therefore possible to make choices when organizing a course. Statements are presented according to their role and their importance in the theory, using the classic terminology: theorems, propositions, corollaries and lemmas. This terminology can be defined as follows: a *theorem* is an important result which is significant for the theory, a *proposition* is less prominent, a *corollary* is a result which can be deduced directly from a major result, a *lemma* is a result with a certain technical nature, which usually is used to prove another result. The end of the proof of a statement is marked by the symbol □.

Graphs, as already mentioned, are used a lot in applications. Many problems are thereby solved in a constructive manner, which leads to writing algorithms, which may then be written in an appropriate programming language. Some of these algorithms are not simple, so it is important to express them in the best structured way possible, which will make the process of writing them in a program easier. In this work we wanted not to neglect, and even to make explicit, this algorithmic aspect, and that is why we cover in detail in Appendix A the method used to express algorithms. Similarly, in Appendix B, we give the bases of what it has become essential to know today for any scientific development, that is the theory of algorithmic complexity. It is impossible to discuss an algorithm without discussing its complexity! However, this assumes at least a good knowledge of the basic classes of the theory of complexity, which are therefore presented in Appendix B. Experience shows that it is still difficult to count on what should normally have been learned elsewhere.

There is today an important body of literature on graphs and their applications. We give only some bibliographic references here, and the reader will find more complete bibliographies within the works cited in the references. Some are relatively old references but they are works that are still interesting today and which have fed our own thoughts in their time and to which we are indebted. We want to recommend in particular Berge [2], a work which, with some others by the same author, is still today a principal reference in graph theory. Some more specific references are also given in the course of the text.

We wish to thank Nicolas Thiant for his very useful contribution to the ultimate verification of the French manuscript. Thanks also to Anne Picard-Pieter and Amit Pieter for undertaking the difficult and delicate task of translating the original French version. In addition, we thank Rebecca Edge for meticulously performing the final proofreading. Finally, we are very grateful to Guillaume and Julie Fournier who have so kindly contributed to finalizing the French and English manuscripts.

Jean-Claude Fournier

Chapter 1

Basic Concepts

After an introduction to the origin of graph conceptualization, this first chapter presents the definitions and basic properties of graph theory. This will be continued a bit later (Chapter 4) with directed graphs. Since graphs are used in many areas, in particular in computer science, the definitions and sometimes the terminology may vary in the literature. In a sense, graphs are victims of their own success! The definitions and terminology chosen here are what seem the most appropriate and the most in line with theory tradition (specifically references [1] and [2]). However, the reader is warned that he may find some differences. We have to know this even if it has no real practical consequences because the concepts remain, of course, the same.

1.1 The origin of the graph concept

Dots and lines linking some dots on a plane: such is the "naïve" view of a "graph". This is already enough to imagine almost everything that this concept can represent in various areas. One example that first comes to mind is a communication network: dots representing the centers or nodes of the network, lines representing the links. Many questions can be asked with this network "model", for example: does this network make it possible for all centers to communicate with one another? In other words, is it possible to reach any given center from another one, directly or going through intermediate centers? Again, given two centers, find the paths linking them, and in particular the "shortest path", that is, the one with the lowest number of intermediary links. This last question may be widened when a numerical value is associated with the link showing a distance between points, or a time, a cost, etc. The length of a path will then consist of the sum of the values of the lines composing it. The shortest path will then be the shortest length path among those linking the two dots being considered. This gives an image of what is called a "weighted graph".

This first example of modeling a communication network with a graph is an application that is particularly obvious. There are other problems which can be modeled with a graph but in a less obvious way. For example, let us consider this rather amusing little problem: a ferryman has to take a wolf, a goat and a cabbage across a river. He can only take one of them across at a time and cannot leave the wolf and the goat alone on one bank of the river (for obvious reasons), nor the goat and the cabbage. How can he proceed to take all of them across? Is there more than one way to solve the problem? Is there one way which is faster than the others? Since there are only a few possibilities, it is quite easy to answer these questions, but it might be interesting to think about similar but more serious problems and try to apply a general and systematic research method. This will be based on modeling the problem with a graph, which is not really difficult but requires a little more imagination than with the communication network (this model is proposed as an exercise at the end of this chapter).

As we go further through this study, there are many practical cases which can be modeled with graphs. It is from some of these cases that "graph theory" emerged and was developed. An 18th century problem, the well-known Königsberg bridge problem (presented in Chapter 9), which was solved by the mathematician Euler, is often mentioned. However, historically, the problem which was the main impetus for the conceptualization and theoretical development of the idea of graphs was the famous "four-color problem". It dates back to the mid-19th century when an English student named Guthrie noticed that it was possible to color a map of his country with only four colors, while respecting the condition that two neighboring regions, that is regions sharing a border (not reduced to isolated vertices), be given two different colors. He wondered if this was a general property, that is, if any geographical map could be colored with only four colors in such a way that two adjacent regions are of different colors. It is quite easy to see that four colors are necessary with the theoretical map shown on the left part of Figure 1.1, which has four regions, pairwise neighboring.[1]

[1]It might be thought that a map with *five* regions pairwise neighboring would require five colors. In fact such a map does not exist because one region would be enclosed by three others and could not border the fifth one. However, this does not solve the four-color problem!

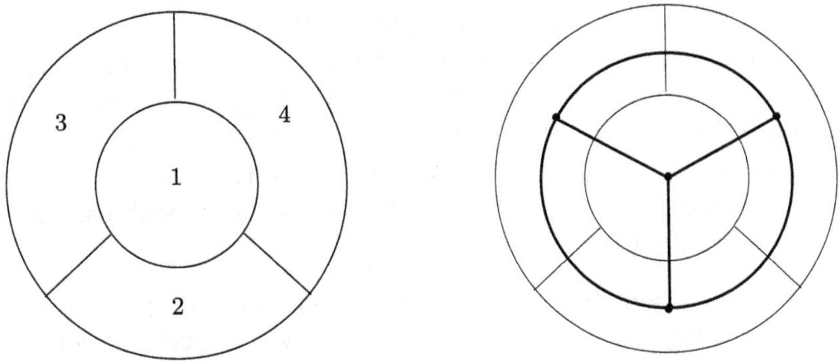

Figure 1.1. *On the left: a map in which the four regions 1, 2, 3, 4 are pairwise neighboring. On the right: the associated graph (in bold)*

This problem, which is really more curious than practical (cartographers have more constraints than giving different colors to neighboring regions and in addition are not limited to four colors), had a fabulous destiny. At the end of the 19th century an amateur mathematician, Kempe, gave what he thought to be a proof and which was acknowledged as such ... until a few years later when another mathematician, Heawood, showed that Kempe's "proof" was false, but that nevertheless we could conclude from it that five colors were always enough! At the time, reasoning with geographical maps in general was not a practical method, and Heawood had to build a map with about 30 regions to show that Kempe was wrong. The error made by Kempe was relatively subtle and the example given by Heawood contradicted Kempe's reasoning but not what was already called the "four-color theorem", which is in fact true as shown in 1976. The manner in which a graph is associated with this problem is simple: a dot in each region and a line linking those dots when the corresponding regions are neighbors, this line goes once through the border line (see the right part of Figure 1.1). First, this graph is "planar", which means that it can be drawn in the plane without two lines intersecting, except for the dots they are linking. Then, the problem of coloring the map is equivalent to giving colors to the various dots of the graph, so that two dots linked by a line are given different colors. The four-color theorem states that any planar graph may have its vertices "colored" in that fashion with no more than four colors. Stated as a graph problem, it was more suitable for the increasingly complex reasoning that was developing. This is how graph language imposed itself during the 20th century. This theorem was only proven in 1976. It was significant news at the time because part of the demonstration was based on the verification of around 2,000 configurations. This verification could only be made on very large computers available at that time and at the cost of more than 1,000 hours of calculation. Since then, the time has been reduced but not enough to do all this work by hand. This was, and still is, a striking, although not unique, example of a mathematical demonstration which called for the use of computers.[2]

We are now going to formalize the concept of a graph; the dots are going to be called "vertices" and the lines "edges". This first concept will be completed later (Chapter 4) with "directed graphs", where the edges are given a direction. In fact, there are many applications which are related to the directed graph model. For instance, in some communication networks the links may only be used in a predefined direction, such as in a town map with one way streets. Another particularly interesting example, to be dealt with later on, is the modeling of the scheduling of a project composed of tasks depending on one another through time, represented in what is called the *potential task graph* (see Chapter 6, section 6.4).

1.2 Definition of graphs

An *(undirected) graph* G is defined by two finite sets:[3] a non-void set X of elements called *vertices*, a set E (which can be empty) of elements called *edges*, with for each edge e two associated vertices, x and y, distinct or not, called the *endvertices* of e.

1.2.1 Notation

We write $G = (X, E)$. The sets X and E can be denoted by $X(G)$ and $E(G)$. The cardinality of X, that is the number of vertices, is usually denoted by n or n_G. The cardinality of E, that is, the number of edges, is denoted

[2]Readers interested in the four-color theorem and its history should consult in particular: *The Four-color Problem, Assaults and Conquest* by T. Saaty and P. Kaimen, McGraw-Hill (1977). A special issue of the *Revue du Palais de la Découverte* (n°12, January 1978, Paris) also deals with this subject and includes the transcript of a conference given at the time by the author of the present work.

[3]Graphs considered in this book are finite.

by m or m_G. The pair of endvertices of an edge e is simply denoted by xy (or yx) instead of the customary mathematical notation $\{x, y\}$ when $x \neq y$.

1.2.2 Representation

It is both usual and practical to draw a graph in a plane in the following manner: the vertices are represented by dots and the edges by simple lines (which can be mathematically defined with precision) connecting two endvertices (see Figure 1.2).

There are, of course, many possible representations for a particular graph because of the different ways there are to place the vertices dots and to trace the edge lines in the plane. For certain applications it is important to find a drawing which shows the structure of the graph, or some of its properties, in relation to the application studied. One example of this is the representation on a screen of the potential task graph in scheduling tasks (discussed in Chapter 6).

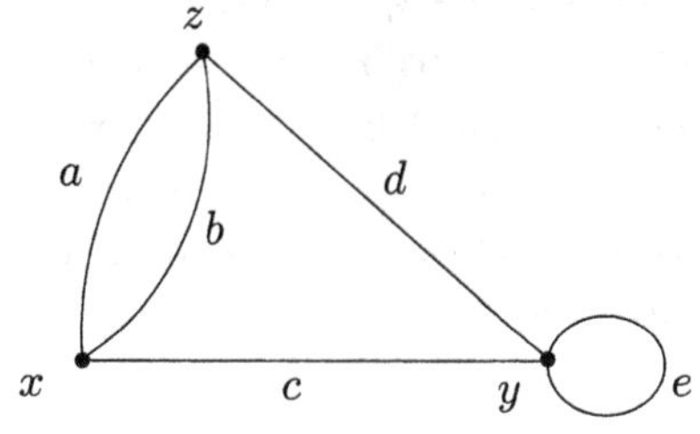

$X = \{x, y, z\}$, $E = \{a, b, c, d, e\}$

Figure 1.2. *A graph and its representation. Edge a is associated with the endvertices x and z. Edge b is also associated x and z. Edge c is associated with x and y. Edge d is associated with z and y. Finally, edge e is associated twice with the vertex y (this is the case for a loop)*

1.2.3 Terminology

When x and y are the endvertices of an edge e, we say that the vertices x and y are *joined* by e, vertices x and y are *adjacent*, or *neighbors*, edge e is *incident* to vertex x and to vertex y. It is possible to have $x = y$; in such a case the edge e is called a *loop*. Two edges e and e', or more, may have the same endvertices x and y; they are said to be *parallel* or that there is a *multiple edge* (*double*, *triple*, etc., depending on the number of edges) joining x and y.

A graph is said to be *simple* if it has no loops or multiple edges. In this case, which often occurs, each edge is identified by its pair of endvertices, which are different, and is denoted for example by $e = xy$. A simple graph may be defined in a simpler way than graphs in the general way previously described as a couple of finite sets (X, E), where X is not empty and E is a set of two subsets of X.

A non-simple graph G is sometimes associated with what is called the *underlying simple graph*, defined as follows: it has the same set of vertices as G and two vertices are joined by an edge if and only if they are different and joined by at least one edge in G.

1.2.4 Isomorphism and unlabeled graphs

We define an *isomorphism* of a graph $G = (X, E)$ to a graph $H = (Y, F)$ by two bijections: ϕ from X on to Y and ψ from E on to F, so that for $e \in E$ and $x, y \in X$, the edge $\psi(e)$ has for endvertices $\phi(x)$ and $\phi(y)$ in H if and only if the edge e has x and y as endvertices in G. This means that these mappings preserve the incidence relation of the edges to the vertices. The graphs G and H are said to be *isomorphic* (see an example of this in Figure 1.3).

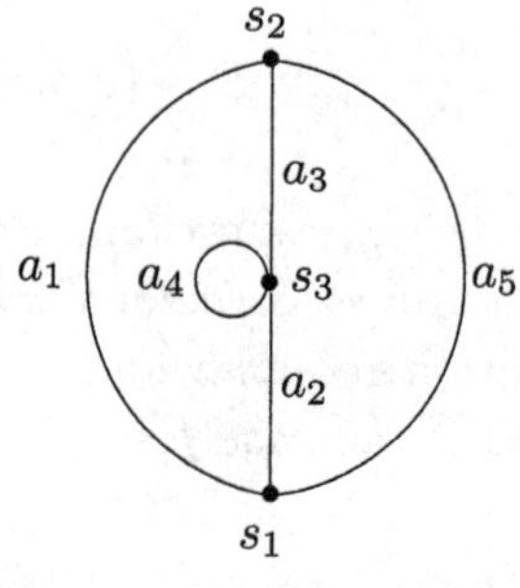

Figure 1.3. *A graph isomorphic to the one in Figure 1.2. The bijections defining the isomorphism between these two graphs can be easily found: vertex s_3 is necessarily associated with vertex y of the previous graph because of the loop a_4, which has to be associated with e in the previous graph. The pair of vertices s_1, s_2 has to be associated with the pair x, z, for example s_1 with x and s_2 with z (a mapping on the edges can be deduced from this)*

Two isomorphic graphs are in fact identical in their graph structure. They have exactly the same properties. They can only be distinguished by the sets of their elements, vertices and edges, in more concrete terms, by the names or labels given to these elements. When focusing solely on the properties of graphs as such, it is natural to consider two isomorphic graphs as similar. This is what we do here, always considering the graphs as *unlabeled.* In other words, in algebraic terms, an unlabeled graph is an equivalence class following the equivalence relation defined on the set of graphs by the isomorphism relation.

1.2.5 Planar graphs

A special and remarkable type of graph is one where it is possible to impose on a plane representation of the graph the condition that two edge lines do not cross each other, except in their common endvertices (in the case of edges having a common endpoint).

This defines *planar graphs*, which play an important role in the theory because of their remarkable properties. We will come back to this point in a later chapter (Chapter 11). Figure 1.4 gives an example.

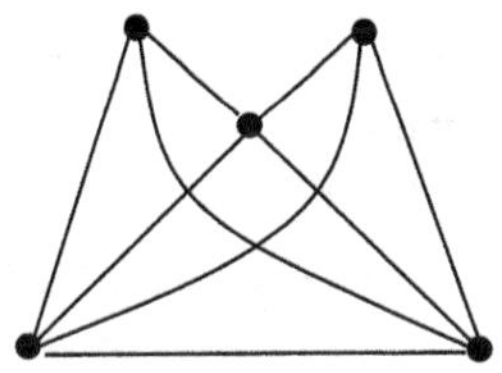

Figure 1.4. *A planar graph not represented in a planar way, since in this representation some edges cross not just at their endvertices. It is easy to find an appropriate representation*

Planar graphs have played a key role for the theory in the famous four-color theorem mentioned above. In graph terms, the theorem states that any planar graph may have its vertices colored with four different colors in such a way that two vertices joined by an edge are of different colors.

1.2.6 Complete graphs

Complete graphs are simple graphs such that any two vertices are joined by an edge. As an unlabeled graph, a complete graph is simply determined by the number n of its vertices. It is generally denoted by K_n (see Figure 1.5 for the case $n = 5$). The number of edges m of K_n is equal to the binomial coefficient $\binom{n}{2}$, that is:

$$m = \frac{n(n-1)}{2}$$

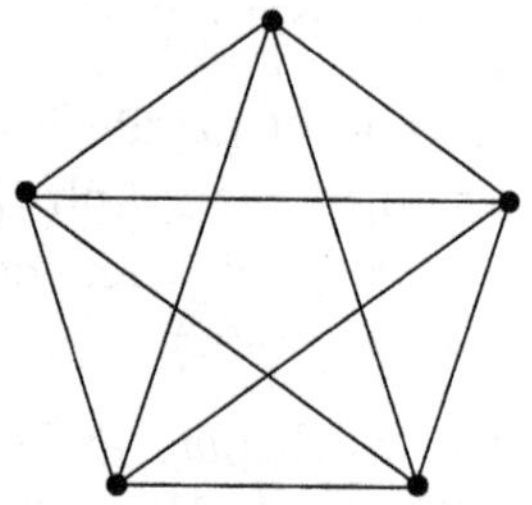

Figure 1.5. *The complete graph K_5*

† NOTE. More generally, we can write for any *simple* graph with n vertices and m edges (indicate why):

$$m \leq \frac{n(n-1)}{2}$$

1.3 Subgraphs

Take the graph $G = (X, E)$. A *subgraph* of G is a graph of the form $H = (Y, F)$, where $Y \subseteq X$ and $F \subseteq E$ are such that any edge of F has its endvertices in Y. Note that the fact that subgraph $H = (Y, F)$ is a graph implies the property that all edges of F have their endvertices in Y.

A subgraph H of G is said to be *induced*, and we can specify *by a set of vertices* $Y \subseteq X$, if it is a graph of the form $H = (Y, F)$, where F is the set of the edges of E whose endvertices are in Y. This subgraph is denoted by G_Y. In particular $G_X = G$.

A subgraph $H = (Y, F)$ of G is called a *spanning subgraph* if $Y = X$. It can be specified that it is a spanning subgraph *induced by* F. It is the graph (X, F) and is denoted by $G(F)$.

Figure 1.6 gives examples of subgraphs.

1.3.1 Customary notation

- $G - Y$, where $Y \subset X$: subgraph of G induced by $X \setminus Y$ (subgraph obtained by removing from G the vertices of Y with their incident edges).
- $G - F$, where $F \subseteq E$: spanning subgraph of G induced by $E \setminus F$ (spanning subgraph of G obtained by removing the edges of F).
- In particular, we will write: $G - x$ in place of $G - \{x\}$ for $x \in X$ and $G - e$ in place of $G - \{e\}$ for $e \in E$.

Expanding from the last case, we sometimes denote by $G + e$ the graph obtained by adding to G a new edge e (mentioning of course its endvertices in G).

1.4 Paths and cycles

1.4.1 Paths

A *walk* of a graph $G = (X, E)$ is a sequence of the form:

$$\left(x_0, e_1, x_1, \ldots, e_k, x_k\right)$$

where k is an integer ≥ 0, x_i are vertices of G, and e_i are edges of G such that for $i = 0, \ldots, k-1$, x_i and x_{i+1} are the endvertices of e_{i+1}. The vertices x_0 and x_k are the *ends* of the walk, and we say that they are *linked* by the walk. The integer k is the *length* of the walk. A walk may have zero length. It is then reduced to a sequence containing only one vertex. When G is a simple graph, a walk may be simply defined by the sequence $(x_0, x_1, \ldots, x_k)$ of its vertices. A *subwalk* of a walk is a walk defined as a subsequence between two vertices of the sequence defining the walk being considered.

A walk is called a *trail* if its edges e_i, for $i = 1, \ldots, k$ are all distinct. We say that the walk *does not go twice* through the same edge.

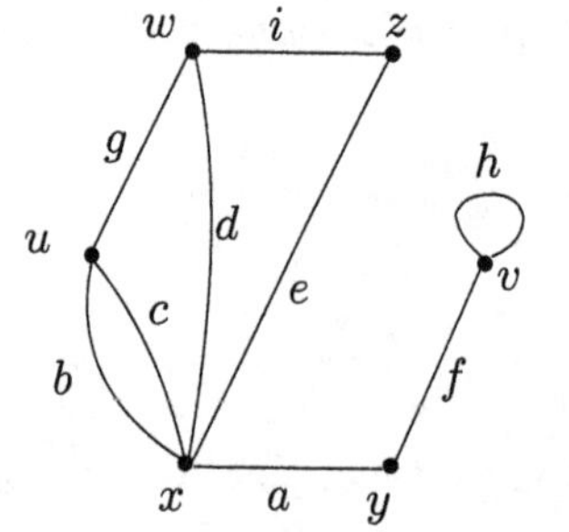

G

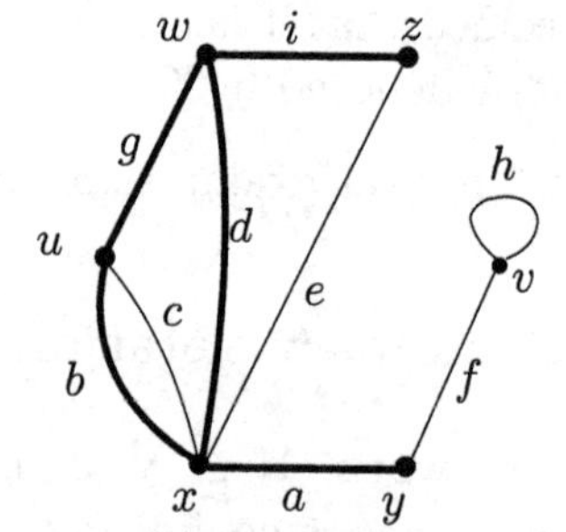

walk:
$(y, a, x, d, w, g, u, b, x, d, w, i, z)$

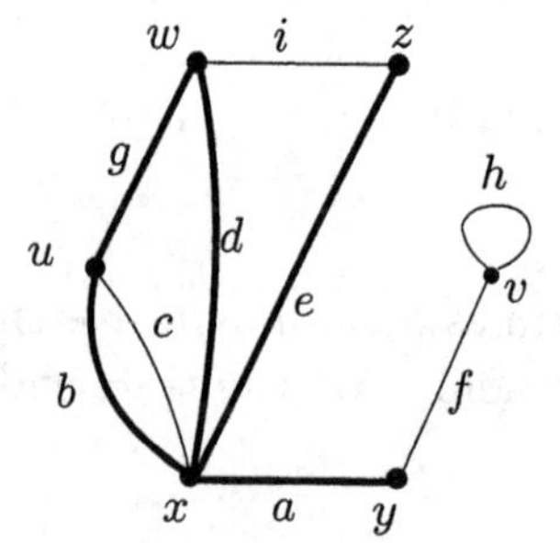

trail:
$(y, a, x, d, w, g, u, b, x, e, z)$

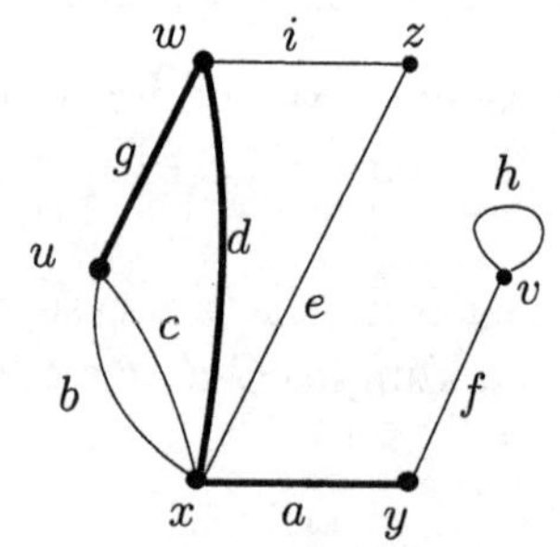

path:
(y, a, x, d, w, g, u)

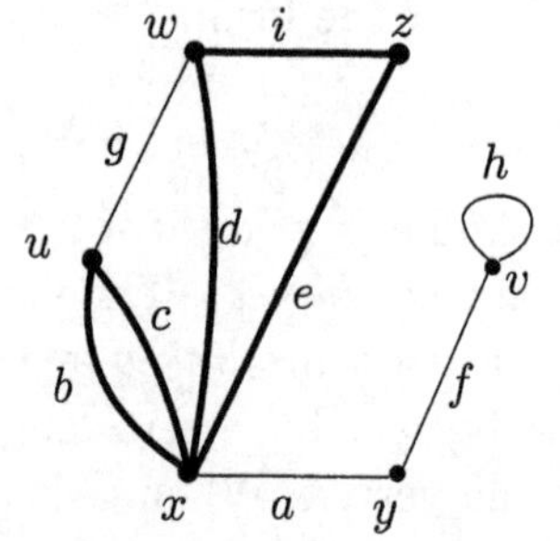

closed trail:
$(x, e, z, i, w, d, x, b, u, c, x)$

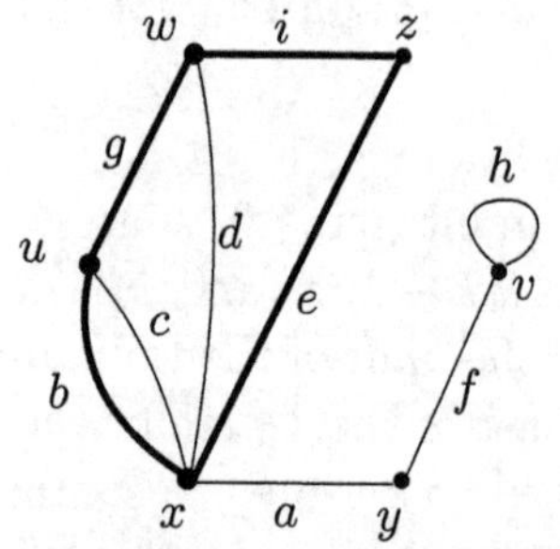

cycle:
$(x, e, z, i, w, g, u, b, x)$

Figure 1.6. *Examples of subgraphs*

A walk is called a *path* if its vertices x_i, for $i = 0, 1, \ldots, k$ are pairwise distinct. It should be noted that a path is necessarily a trail.

The following result is often useful for reasonings where walks are concerned.

LEMMA 1.1. *In a graph, if two vertices are connected by a walk then they are connected by a path.*

Proof. Given a walk linking the vertices x and x' of a graph G and in which one vertex appears twice:

$$(x = x_0, e_1, x_1, \ldots, x_i, e_{i+1}, \ldots, x_j, \ldots, e_k, x_k = x')$$

where $x_i = x_j$ with $0 \leq i < j \leq k$. The walk can be shortened by removing the subsequence (subwalk) between x_i and x_j, which gives a new walk still linking x and x':

$$(x = x_0, e_1, x_1, \ldots, x_i = x_j, \ldots, e_k, x_k = x')$$

By repeating this shortening process as long as there is a vertex that can be found twice in the walk, that is as long as the walk is not a path, we end up obtaining a path linking the vertices x and x'. □

1.4.2 Cycles

A walk, a trail or a path $(x_0, e_1, x_1, \ldots, e_k, x_k)$ is said to be *closed* if its ends x_0 and x_k coincide.

A *cycle* is a closed path of length ≥ 1, that is a path of the form:

$$(x_0, e_1, x_1, \ldots, e_k, x_0)$$

where $k \geq 1$, and vertices x_i, for $i = 0, \ldots, k-1$, are all distinct. Integer k is the *length* of the cycle. Unlike a walk, and a trail, a cycle cannot have zero length. The minimum length it can be is 1. In this case, it is made up of one vertex with a loop. When G is a simple graph, a cycle may be defined by the sequence $(x_0, x_1, \ldots, x_0)$ of its vertices. In this case the length is the number of vertices of the cycle (except the last one). A cycle is called *even* or *odd*, depending on whether its length is even or odd.

Figure 1.7 gives some examples of walks, trails, paths and cycles.

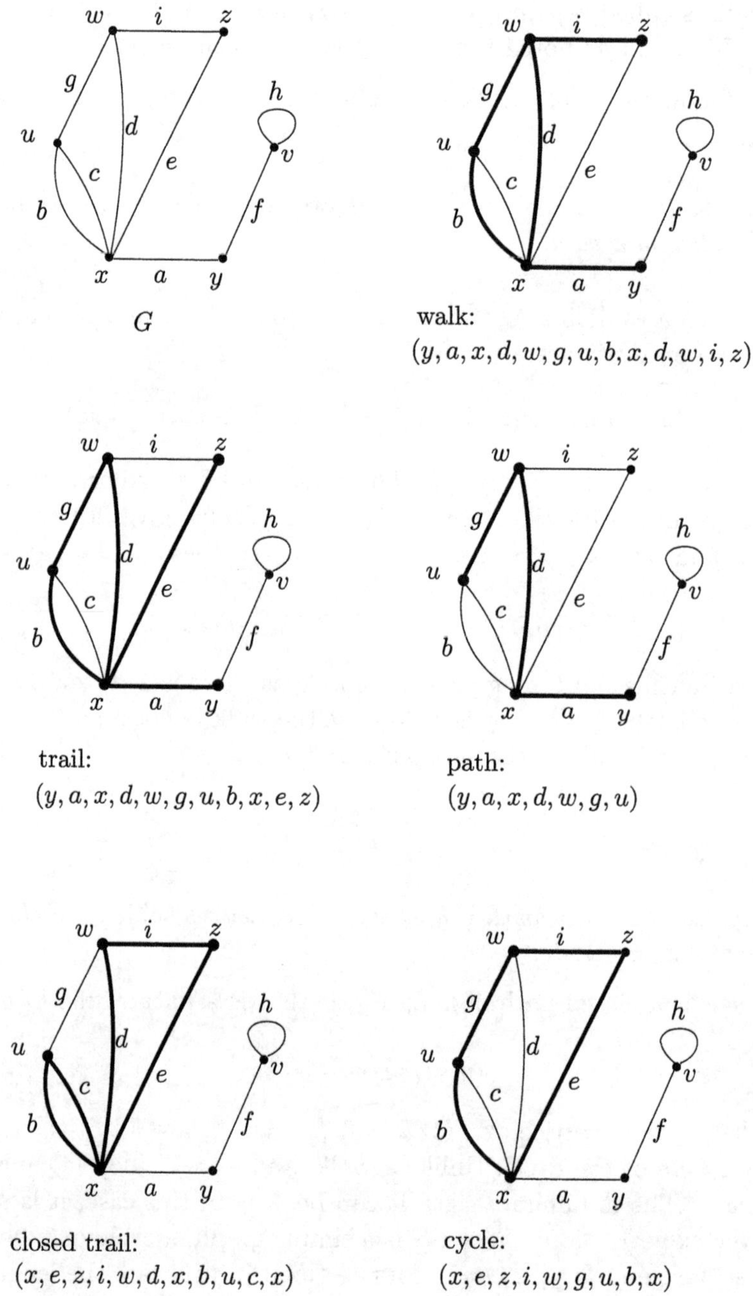

Figure 1.7. *Examples of walks, trails, paths, and cycles*

NOTE. No distinction is made between cycles that only differ in the cyclic sequences of vertices which define them. For example, in a simple graph, the three following cycles are considered as one unique cycle:

$$(x_0, x_1, x_2, x_3, x_4, x_0)$$
$$(x_3, x_4, x_0, x_1, x_2, x_3)$$
$$(x_0, x_4, x_3, x_2, x_1, x_0)$$

In fact, in the graph, it is the same cycle described differently.

1.4.3 Paths and cycles as graphs

By extension of the preceding definitions, a *path* is also a graph $G = (X, E)$ where $X = \{x_0, x_1, \ldots, x_k\}$ and $E = \{e_1, \ldots, e_k\}$ so that:

$$(x_0, e_1, x_1, \ldots, e_k, x_k)$$

is a path of G. It is in this sense that we will say that a path is a tree (see next chapter).

In the same way, we can call a *cycle* a graph which can be described like a cycle. Taking into account what has been said above and the fact that we are dealing here with unlabeled graphs, such a cycle is unique as long as its length is fixed. That is why we can say *the* cycle of length 5.

1.5 Degrees

The *degree* of a vertex x in a graph G is the number of edges in G incident to x, that is edges with x as an endvertex, loops being counted twice. This integer is denoted by $d(x)$ or $d_G(x)$. For example, for the graph in Figure 1.2: $d(x) = 3$, $d(y) = 4$, $d(z) = 3$.

A vertex is *isolated* if its degree equals zero. Dealing with degrees is an opportunity to state the following proposition.

PROPOSITION 1.1. *In any graph G, we have:*

$$\sum_{x \in X} d_G(x) = 2m.$$

Proof. When adding up the vertex degrees of G, each edge is counted twice, once for each end (this is particularly true with loops since each loop counts twice in the degree). The result is thus twice the number of edges of the graph. □

The method for this proof is a little like counting a herd of sheep: let us count the legs and divide the result by four (although for sheep there is always the question of five-legged sheep!). The following corollary, when applied in a different context from graphs, may appear far from self-evident.

COROLLARY 1.1. *In a graph the number of vertices with odd degrees is even.*

Proof. The sum of the degrees being even, since it is equal to twice the number of edges, can only include an even number of odd terms. Therefore, there is an even number of odd degrees in the graph. □

Here is an amusing application of this corollary. Let us imagine a group of nine friends who either shake hands or give each other hugs as a greeting in the morning. Each one of them shakes the hand of three of his friends and hugs the other five. This is in fact impossible! Let us model the situation of these friends by a graph which could be called "the hugging graph": the vertices are the friends and two vertices are linked by an edge if and only if the related friends greet each other with a hug (this is a *simple* graph, in particular because no friend is assumed to greet himself). Any vertex is of degree 5 and there are nine vertices. This contradicts the preceding corollary.

The *minimum degree* of a graph G is the smallest degree of its vertices and is denoted by δ_G or simply δ. It should be observed that δ_G is the degree of at least one of the vertices of the graph. Likewise the *maximum degree* of G is the largest degree of its vertices and is denoted by Δ_G or simply Δ. This is also the degree on at least one of the vertices of the graph.

NOTE. The following inequalities result from proposition 1.1:

$$n\delta_G \leq 2m \leq n\Delta_G$$

1.5.1 Regular graphs

A graph G is said to be *regular* when the degrees of its vertices are all equal. The common degree, say k, may be specified by calling it a *k-regular* graph. A 3-regular graph is said to be *cubic*. Cubic graphs have many properties and have played a key role in the four-color theorem.

NOTE. We have in a k-regular graph G with n vertices and m edges:

$$nk = 2m$$

This useful formula can be directly deduced from proposition 1.1.

1.6 Connectedness

A G graph is said to be *connected* if any two vertices of this graph are linked by a path in G. Otherwise, the graph is *a disconnected graph.*

The *connected* components of a graph G are the maximal connected induced subgraphs of G. *Maximal* means here that the subgraph mentioned is not itself a proper subgraph, that is with strictly fewer vertices, of a connected subgraph of G. Obviously, a graph is connected if and only if it has only one connected component.

We verify that the connected components of a graph are subgraphs pairwise disjoint, that is having pairwise no common vertices and no common edges. It defines *the decomposition into connected components* of the graph (see Figure 1.8 for an example). This decomposition is unique.

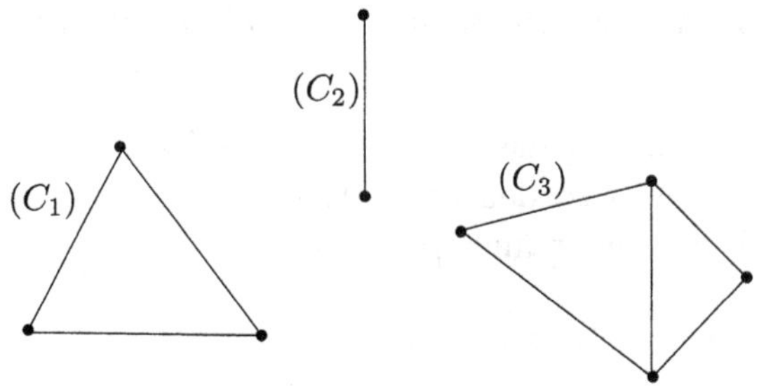

Figure 1.8. *A disconnected graph and its three connected components:* C_1, C_2, C_3

It is also possible to define in algebraic language the connected components of a graph $G = (X, E)$ as the subgraphs induced by equivalence classes over X, defined by the relation: the vertices x and y are linked by a path. This binary relation is in fact an equivalence relation on the set X (reflexive, symmetric and transitive).

To finish connectedness, let us just mention the following proposition: *If a graph possesses a spanning subgraph which is connected, it is itself connected.* This proposition is one of many small propositions which are often not proved or even stated. Nevertheless it is useful for a beginner in † graph theory to practice by proving them rigorously at least once. If we can do this easily, then all is well, at least so far into the theory. If we do not succeed, we should go back over the preceding pages or maybe rethink our personal logic.

1.7 Bipartite graphs

A graph G is *bipartite* if the set of its vertices can be divided into two disjoint subsets such that each edge has an endvertex in each subset. We denote a bipartite graph by $G = (X, Y, E)$, where X and Y are the two subsets of vertices (and so $X \cup Y$ is the set of all vertices) and E is the set of edges.

NOTES. 1) It is important to note that one of the sets X or Y can be empty. As a result, the couple (X, Y) is not mathematically, strictly speaking, a partition (the sets of a partition should not be empty). Nevertheless the terms "bipartition" and "classes" are often used. It should be noted that with this definition a graph reduced to one vertex, and no edge, is bipartite.

2) A bipartition which defines a graph as bipartite is generally not unique.

3) A bipartite graph has no loops. Indeed a loop would contradict the hypothesis that an edge has its endvertices in different sets. However, a bipartite graph may have multiple edges.

A bipartite graph $G = (X, Y, E)$ is *complete* if it is simple and the set of its edges is $E = \{xy \mid x \in X,\ y \in Y\}$, that is any pair of a vertex of X and of a vertex of Y is an edge of G. It is denoted by $K_{p,q}$, where p is the cardinality of X and q the cardinality of Y (see Figure 1.9 for an example).

Bipartite graphs are important in graph theory and for certain applications (for example matchings, dealt with in Chapter 7). They are also interesting as they can be easily characterized by a property of cycles as in the following classic result.

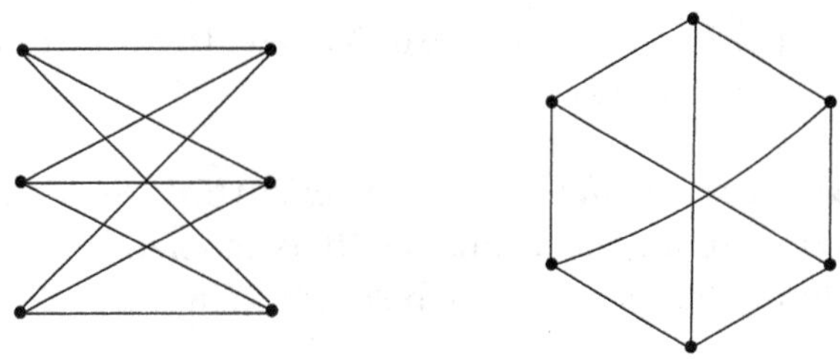

Figure 1.9. *Two ways of representing the complete bipartite graph $K_{3,3}$*

1.7.1 Characterization

THEOREM 1.1. *A graph is bipartite if and only if it contains no odd cycle.*

Proof (outline). The proof of the necessary condition is easy when reasoning by the absurd and, in relation to the classes of the bipartition, following in order the vertices of an odd cycle. The proof of the sufficient condition is less simple but can be done in a constructive way, that is by producing the adequate bipartition. The principle is as follows: mark a first arbitrarily chosen vertex 0, then mark its neighbors 1, then take each of the newly marked vertices and mark their not-yet-marked neighbors 0, and so on until all vertices reached are marked 0 or 1. The crucial point is that if during this marking process two neighboring vertices happen to receive the same mark (twice 0 or twice 1) then there is an odd cycle in the graph. This can be seen by considering the paths defined by the succession of marked vertices which come to these two vertices and the edge joining them. With the hypothesis of the sufficient condition, this circumstance of two neighboring vertices bearing the same mark will not occur. The marks given to the vertices will define a bipartition in compliance with the definition of bipartite graphs. Any vertex will be marked as soon as the graph is connected, otherwise we should proceed independently with each connected component. □ †

1.8 Algorithmic aspects

Graph theory is a field of great relevance to algorithmic studies. Many applications involve algorithms of graphs and the graphs being considered must be represented for the computer. Furthermore, the efficiency of the algorithms must be evaluated. This question is related to complexity theory (see Appendix B), the bases of which are assumed to be known to the reader. Let us consider a few problems:

- *To find out if two graphs are isomorphic*: this problem, though basic, does not yet have any clear answer. It is in class **NP**, but not known as **NP**-complete neither does it belong to class **P**.

- *To find out if a graph is connected*: this problem is solved linearly. Classical searches of graphs such as the *depth-first search* described in Chapter 5 address this question.

- *To find out if a graph is planar*: this problem is solved linearly. It was recognized early on as a class **P** problem. However, its solution by a linear algorithm was much more difficult to obtain (this was achieved in the 1970s).

- *To find out if a graph is bipartite*: this problem is solved linearly. The proof of the sufficient condition of theorem 1.1 outlined above describes a process which leads to a linear algorithm for recognizing a bipartite graph. This algorithm is based on a *breadth-first search* (described in Chapter 6).

1.8.1 Representations of graphs inside a machine

There is no unique answer to the implementation of graphs in computers. On the one hand it is possible to imagine many models *a priori* but, on the other hand, the implementation of a graph depends on the way in which it will be used. The choice of a model may have a direct influence on the efficiency of the algorithm in terms of complexity. In order to classify the various computer models of graphs, the three following principles can be distinguished:

1. Give the possibility of finding out if two given vertices are neighbors, that is joined by an edge in the graph. Furthermore, in the case of a non-simple graph, give the number of edges joining the vertices under consideration. The natural way to do this for a graph $G = (X, E)$ is the *adjacency matrix* defined as follows: setting $X = \{x_1, \ldots, x_n\}$, this is the square matrix of order n, $M = (m_{ij})$, where m_{ij} is the number of edges having x_i and x_j for endvertices in G. Such implementation of a graph takes memory space of the order n^2, where n is the number of vertices of the graph. Taking into account that processing the graph

takes at least the time needed to read its data, this means that any algorithm over graphs modeled after an adjacency matrix will require a time complexity of at least $O(n^2)$. Nevertheless, as we will see later, for some algorithms the processing in itself has a complexity of lesser order, for example $O(n)$, therefore this model is not always appropriate.

2. A second principle for implementing a graph is to give for each vertex its "neighborhood": the vertices which are its neighbors or its incident edges or even both, that is its incident edges and their endvertices. This last method will be particularly useful when there are multiple edges. Implementing this can be done in various ways, the most classic, from a programming point of view, is to give, for example, the neighbors in a list called an *adjacency list* or *list of neighbors*. If we want to modify the graph during its processing, it is best to implement these lists under the classic form of *linked lists*. Data in lists of neighbors makes it possible to search one after the other for the neighbors of each vertex, which is a classic situation for algorithms of graphs (for example in Chapter 5 for a depth-first search of a graph). Sometimes it is possible to take a less structured processing approach and give simply the *set* of the neighbors $N(x)$ for each vertex x (for instance in Chapter 7 to research matchings). The memory space necessary for this type of implementation is of the order $n + m$, where n is the number of vertices and m the number of edges of the graph. Such a model is coherent with linear processing.

3. If in an algorithm the entry to the graph is done by the edges and not directly by the vertices, as in the above models, a third implementation principle consists of giving a *list of edges* with, for each of them, its endvertices (see such an example in Chapter 2 with Kruskal's algorithm). Such a list may be *linked*, which makes it possible to modify the graph being processed. The memory space necessary for this type of model is minimal since it is of the order of the number of edges m of the graph (note that the isolated vertices do not appear in this implementation, since they are not endvertices of an edge, by definition).

Figure 1.10 shows, for a simple graph, these models in some specific cases of the three principles above. Adjacency matrices and list of edges, as arrays, can be understood easily. Lists of neighbors are drawn as *linked lists*. The principle of a linked list, a classic structure in algorithms and programming,

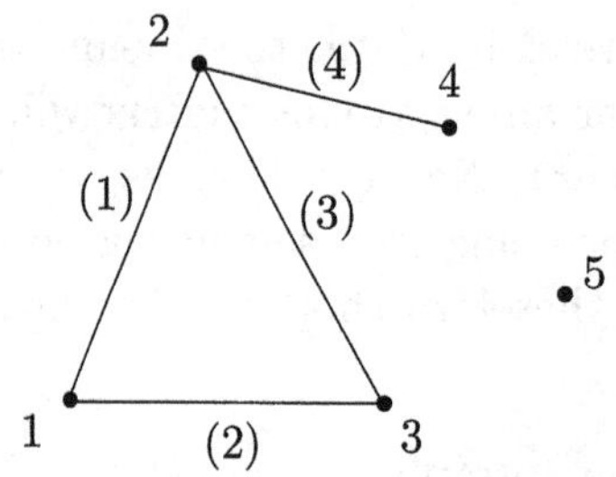

graph

	1	2	3	4	5
1	0	1	1	0	0
2	1	0	1	1	0
3	1	1	0	0	0
4	0	1	0	0	0
5	0	0	0	0	0

adjacency matrix

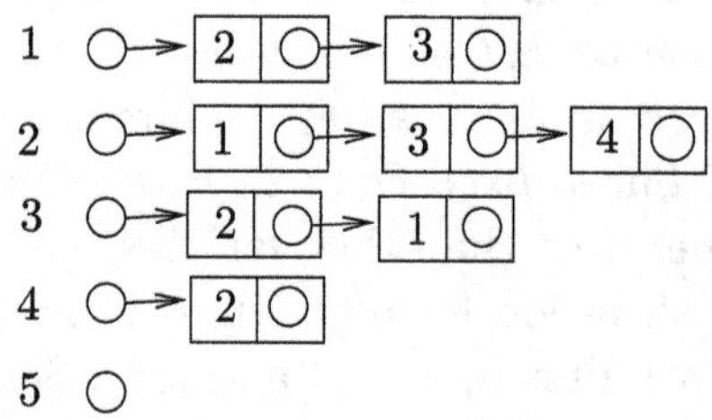

lists of neighbours

edges	endvertices
1	1 2
2	3 1
3	2 3
4	2 4

list of edges

Figure 1.10. *Various representations of a simple graph inside a machine*

is that each item of the list is associated with the address of the next item, an address which is called a *pointer*. Each item of the list is represented by a rectangle with two boxes: in the first one the neighbor is given and in the other the pointer to the following item. The pointer is represented by a circle with an arrow (pointing towards the next item). A circle without an arrow indicates the end of a list or an empty list since there is no following item. For example, we read that the neighbors of vertex 1 are: 2, 3, or that vertex 5 has no neighbor (empty list). Note that in this representation of lists of neighbors, the arrows have nothing to do with the edges of the graph.

These principles are basic and in specific cases more precise information may be necessary. In addition to the above descriptions, more specific models may be defined for some applications.

1.8.2 Weighted graphs

In graph applications, in particular in optimization, *weighted graphs* are often considered, that is graphs with values, integer or real, positive or not, associated with the edges. Formally, we have a graph $G = (X, Y)$ with a mapping $v : E \to \mathbb{R}$.

When a weighted graph is a simple graph, which is often the case, its computer model is generally a matrix, such as the adjacency matrix, but with entries being the values of the edges under consideration. We choose a special number, for example ∞, when there are no edges joining the vertices associated with this entry of the matrix. Specifically, using the above notation, it is the matrix $M = (v(x_ix_j))$, where $1 \leq i, j \leq n$, with mapping v extended by stating: $v(x_ix_j) = \infty$ when $i \neq j$ and $x_ix_j \notin E$, $v(x_ix_j) = 0$ when $i = j$. This matrix is symmetric.

It is also possible to use the list of edges to represent weighted graphs, by adding for each edge x_ix_j the data of its value $v(x_ix_j)$. In practice, it is possible to define an array indexed on the "edge" type of the graph. This type is defined as an interval of integers by numbering the edges from 1 to m, and by associating with each edge a record containing three fields: two for the endvertices of the edge and one for its value.

The list of neighbors is *a priori* less adapted to represent weighted graphs. Nevertheless, it is possible in the case of simple weighted graphs to add for each neighbor the data of the value of the corresponding edge.

1.9 Exercises

1.1. Find two simple connected graphs with the same number of edges and the same number of vertices which are not isomorphic.

1.2. Find all non-isomorphic simple graphs having respectively one, two, three and four vertices (for four vertices there are 11 such graphs).

+1.3. Show that in a graph G, which is not assumed to be connected, for each vertex x of odd degree there is a vertex $y \neq x$ of odd degree such that x and y are linked by a path. Deduce from this that if G has exactly two vertices of odd degree, then these are linked by a path.

+1.4. Show that any closed trail in a graph can be decomposed into edge-disjoint cycles.

1.5. G is a simple graph with n vertices ($n > 1$). Note that since it is a simple graph, the degree of every vertex of G is strictly less than n.

a) Show that it is impossible to have a vertex of degree 0 and a vertex of degree $n-1$ at the same time in G.

b) Deduce from this that there are at least two vertices of the same degree.

(A direct application of this result is to be able to say that in any set of at least two people there are necessarily two of them who have the same number of friends in that set.)

1.6. Show that a graph is connected if and only if there is no bipartition of the set of its vertices such that no edge has an endvertex in each subset of this bipartition.

*1.7. a) Show that if a simple graph G verifies the following inequality

$$m > \frac{1}{2}(n-1)(n-2)$$

then it is a connected graph.

b) Find, for all $n \geq 2$, a simple disconnected graph G such that:

$$m = \frac{1}{2}(n-1)(n-2)$$

*1.8. In a simple graph G, a *triangle* is a triplet of distinct vertices x, y, and z such that xy, yz, and xz are edges of G. $G = (X, E)$ is a simple graph, n the number of its vertices and m the number of its edges. Let us suppose that G contains no triangles.

a) Show that for two distinct neighbor vertices x and y, the number n_x of vertices of $X \setminus \{x, y\}$ neighbors of x and the number n_y of $X \setminus \{x, y\}$ neighbors of y satisfy this inequality:

$$n_x + n_y \leq n - 2$$

b) Deduce from this, by *induction on* n, the inequality:

$$m \leq \frac{n^2}{4}$$

c) Find an infinite family of simple graphs without triangles for which:

$$m = \frac{n^2}{4}$$

+1.9. Show that if a bipartite graph $G = (X, Y, E)$ is k-regular then this graph is *balanced*, that is $|X| = |Y|$ ($|X|$ represents the number of elements of X and similarly for $|Y|$).

1.10. *(About planar graphs)*

a) Show that complete graph K_4 is planar. Draw it with straight line segments as edges.

b) Is complete graph K_5 a planar graph? (The answer is no. How can this be justified?)

1.11. Given M the adjacency matrix of a graph G whose set of vertices is $X = \{x_1, x_2, \ldots, x_n\}$, show that the term (i, j) of M^k, the kth power of matrix M, where k is an integer ≥ 1, is the number of walks which link the vertices x_i and x_j in G.

1.12. *(Going from one computer representation to another)*

Write algorithms of passage between the different representations of a graph: adjacency matrice, lists of neighbors, lists of edges. It is particularly interesting to consider going from a representation by list of edges to one by lists of neighbors. Try to minimize the reading of the list of edges. Analyze the complexity of the algorithm built.

*1.13. Try to build an algorithm to determine the connected components of a graph. Analyze its complexity.

1.14. Let $G = (X, E)$ be a simple graph. The *complement* $\overline{G}$ of G is the simple graph whose vertex set is X and whose edges are the pairs of non-adjacent vertices of G. Show that G or $\overline{G}$ (or both) is connected.

Chapter 2

Trees

In this chapter we will define and study a set of graphs, known as trees, which play a major role in graph theory as well as in its applications. This study will be completed with the study of arborescences in Chapter 4.

2.1 Definitions and properties

A *tree* is a connected acyclic graph, where *acyclic* means without a cycle. A tree is a simple graph (a loop or a double edge would define a cycle).

A path is in particular a tree (see Figure 2.1).

Except when explicitly noted otherwise, as with graphs in general, n denotes the number of vertices of a tree and m the number of edges.

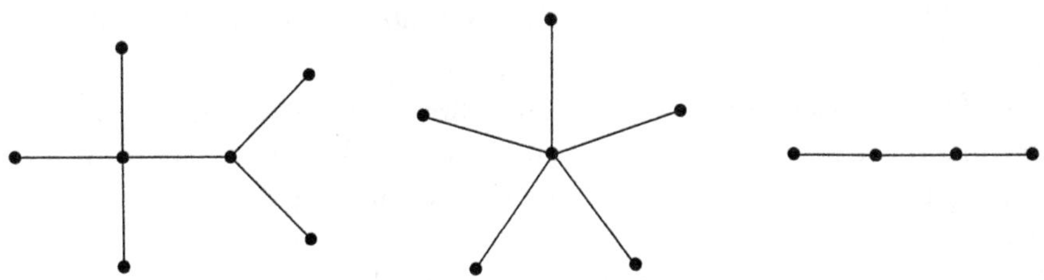

Figure 2.1. *Three examples of trees*

2.1.1 First properties of trees

PROPOSITION 2.1. *A tree such that $n \geq 2$ has at least two vertices of degree one.*

Proof. Consider in a given tree a maximal path, that is, a path not contained in a strictly longer path. Let $(x_0, e_1, x_1, \ldots, e_k, x_k)$ be this path. Since $n \geq 2$, we have necessarily $k \geq 1$. Let us show that the ends of this path, x_0 and x_k, which are distinct, are vertices of degree one in the tree. Suppose that vertex x_0, for example, is not of degree one and thus has an incident edge $f \neq e_1$, and let y be the endvertex of f other than x_0. If vertex y is one of the vertices of the path, given x_j, then there is in the graph a cycle, $(x_j = y, f, x_0, e_1, \ldots, x_j)$, which contradicts the hypothesis that this graph is a tree. If vertex y is not one of the vertices of the path, then this path is not maximal, because it is strictly included in the path $(y, f, x_0, e_1, x_1, \ldots, e_k, x_k)$. In both cases, there is a contradiction. □

PROPOSITION 2.2. *If G is a tree then $m = n - 1$.*

Proof. Apply inductive reasoning to n. For $n = 1$, we have $m = 0$ since an edge could only be a loop, i.e. a cycle, which is impossible for a tree. Assume $n > 1$. According to proposition 2.1 above, there is in G a vertex of degree one, given x. The subgraph $G - x$ is connected since suppressing a vertex of degree one in a connected graph does not suppress the connectedness property, as is easily verified. In addition, $G - x$ is acyclic because a cycle in $G - x$ would also be a cycle of G, while G, as a tree, is itself acyclic. Thus $G' = G - x$ is a tree and it is possible to apply the induction hypothesis, which gives: $m_{G'} = n_{G'} - 1$. As $m_{G'} = m_G - 1$ and $n_{G'} = n_G - 1$, we can deduce the desired equality: $m_G = n_G - 1$. □

PROPOSITION 2.3. *In a tree any two vertices are connected by a unique path.*

Proof. The existence of a path linking two vertices of the graph is direct according to the hypothesis of connectedness. To show that it is unique, suppose the existence of two distinct paths linking two vertices x and y. If $x = y$ then one of these two closed paths is of length ≥ 1 and so is a cycle, which contradicts the hypothesis of the tree being acyclic. If $x \neq y$, the concatenation of the two paths linking x and y is a closed walk. This closed walk is not necessarily a cycle. This is due to the fact that the two paths may share one or more edges (not all of them, since those paths are distinct according to the hypothesis). A small technical work, left to the reader, shows that it is always possible to extract a cycle from the concatenation of the two paths, again a contradiction. □

2.1.2 Forests

A *forest* is an acyclic graph. The connected components of a forest are therefore trees, which explains the use (very natural!) of the terminology. Forests generalize trees.

PROPOSITION 2.4. *In a forest G, we have $m \leq n - 1$, with equality if and only if G is a tree.*

Proof. Given C_1, C_2, ..., C_p the connected components of G, apply proposition 2.2 to each of these components, denoting respectively n_i and m_i the number of vertices and the number of edges of C_i. By adding all these equalities, for $i = 1, 2, \ldots, p$, we then have:

$$\sum_{i=1}^{p} m_i = \sum_{i=1}^{p} (n_i - 1) = \sum_{i=1}^{p} n_i - \sum_{i=1}^{p} 1$$

Thus:

$$m = n - p$$

and since $p \geq 1$ (p is the number of connected components of G):

$$m \leq n - 1$$

Equality occurs if and only if $p = 1$, that is if and only if G is connected. Since G is by hypothesis acyclic, this means if and only if G is a tree. □

2.1.3 Bridges

A *bridge* of a graph G is an edge e such that $G - e$ has one more connected component than G. This is a way of saying that in $G - e$, the endvertices x and y of e are no longer linked by a path. We could also say that the edge e *separates* the vertices x and y. When G is connected, a bridge is an edge e so that $G - e$ is disconnected.

LEMMA 2.1. *An edge of a graph G is a bridge if and only if it does not belong to a cycle of G.*

Proof. It is sufficient to consider the case where the graph G is connected. e is an edge of G, and x and y its endvertices. If e is not a bridge, there is in $G - e$ a path linking x and y. This path constitutes with the edge e a cycle of G. This cycle contains the edge e, which demonstrates the sufficient condition. Assume now that the edge e belongs to a cycle:

$$C = \big(x_0, e_1, x_1, e_2, \ldots, x_{k-1}, e_k, x_0\big)$$

with, for example, $e = e_1$ (which can always be assumed). Let us show that $G - e$ is connected, which means that the edge e is not a bridge of G. u and v are any two vertices of $G - e$. These vertices are linked in G, a connected graph by hypothesis, by a path:

$$D = \big(u = y_0, f_1, y_1, \ldots, f_k, y_k = v\big)$$

If this path does not pass through the edge e, it is also a path of $G - e$. Otherwise, $e = f_i$ for an $i \in \{1, \ldots, k\}$ with, for example, for the endvertices: $x_0 = y_{i-1}$ and $x_1 = y_i$. Substitute in path D edge f_i, which has y_{i-1} and y_i as endvertices, the following path extracted from path C (it corresponds to the path C deprived of the edge e_1 and read in the reverse direction):

$$\big(y_{i-1} = x_0, e_k, x_{k-1}, \ldots, e_2, x_1 = y_i\big)$$

The walk D' thus obtained links the vertices u and v in $G - e$. Thus (lemma 1.1), there is a path which connects the vertices u and v in $G - e$, which ends the proof of the necessary condition and of the lemma. □

Bridges are involved in tree properties through the following result.

PROPOSITION 2.5. *In a tree, any edge is a bridge.*

Proof. This is a direct consequence of the definition of trees and of lemma 2.1 above. □

2.1.4 Tree characterizations

THEOREM 2.1. *The following conditions for a graph G are equivalent:*

(1) G is a tree.

(2) G is connected and $m = n - 1$.

(3) G is acyclic and $m = n - 1$.

(4) G is connected and every edge is a bridge.

(5) In G any two given vertices are linked by a unique path.

Proof. The implications (1)⇒(2), (1)⇒(3), (1)⇒(4), (1)⇒(5), (3)⇒(1) result directly from the above propositions. Implication (4)⇒(1) is straightforward with lemma 2.1. Implication (5)⇒(1) is easy: if there was a cycle in G, one of its vertices would be joined to itself on the one hand by the cycle, considered as a (closed) path, and on the other hand by the path with of zero length that this vertex defines. This contradicts the hypothesis of the uniqueness of a path linking any two vertices. To end the proof, that is to verify that these implications are sufficient, we must demonstrate implication (2)⇒(1). Consider a graph G verifying (2). Remove, as long it is possible, an edge which is not a bridge (first in graph G, and then in the current graph obtained). The spanning subgraph G' obtained is connected, like G, because each of the edges removed was not a bridge. It is also an acyclic graph since it now has nothing but bridges and thus cannot have any cycle (lemma 2.1). This graph G' is therefore a tree, spanning a subgraph of G. Let m' be the number of edges of G'. We have $m' = n - 1 = m$. Thus, G' having the same number of edges as G, $G' = G$ and G is therefore a tree. □

NOTE. The method of proving equivalences between some conditions, which is used here to prove this theorem, is classic. It consists of demonstrating a set of implications which implies all the others. This corresponds to the concept of "transitive closure" of a binary relation, closure which must here be "complete" in the sense that it contains all pairs. This concept can be expressed in terms of directed graphs (see later definition and exercise 4.3 in Chapter 4).

2.2 Spanning trees

A *spanning tree* of a graph G is a spanning subgraph of G which is a tree (see example in Figure 2.2).

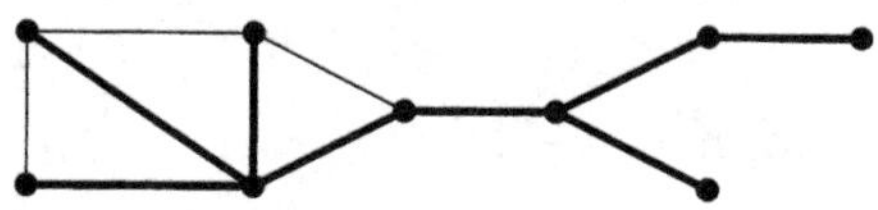

Figure 2.2. *A spanning tree of a graph (in bold)*

PROPOSITION 2.6. *A connected graph G has (at least) one spanning tree.*

Proof. Remove from G, if possible, an edge which is not a bridge. On the one hand, the spanning subgraph obtained is always connected since only non-bridge edges were removed from the current graph. On the other hand, this subgraph has no cycle since it no longer has any bridge. Therefore, it is a tree. By construction, it is also a spanning subgraph of G. □

COROLLARY 2.1. *If G is connected then $m \geq n-1$, with $m = n-1$ if and only if G is a tree.*

Proof. Since G is connected, it contains, as a spanning subgraph, a tree T. We have $m_G \geq m_T = n_T - 1 = n_G - 1$. Since T is a spanning subgraph of G, the equality is possible only if $G = T$, that is if G is itself a tree. □

The two following results will be useful in particular for the application dealt with later on.

PROPOSITION 2.7. *A spanning subgraph of a connected graph G is a spanning tree of G if and only if it is connected and edge-minimal.*

Proof. The necessary condition results from proposition 2.5. In order to prove the sufficient condition, let T be a spanning subgraph of G which is connected and minimal. Then for any edge e of T, $T - e$ is no longer connected, that is, e is a bridge of T. Condition (4) of theorem 2.1, applied to T, allows us to conclude. □

PROPOSITION 2.8. *A spanning subgraph of a connected graph G is a spanning tree of G if and only if it is acyclic and edge-maximal.*

Proof. T is a spanning tree of G. If such an edge exists, let e be one which does not belong to T. The endvertices of e are linked in T by a path (since

T is connected). This path with edge e defines a cycle of $T + e$. Thus, T is really acyclic and maximal, which proves the necessary condition. Now, given T a spanning subgraph of G which is acyclic and maximal, to show that T is a spanning tree, and to justify the sufficient condition, we only need to show that T is connected. Let x and y be any two vertices of T and let us show that they are linked by a path of T. Since G is connected, there is a path D of G linking x and y. If this path has all of its edges in T, we are done. If not, let e be an edge of D which is not in T. By hypothesis on T, $T + e$ has a cycle, C. This cycle contains the edge e, and according to lemma 2.1 this edge is not a bridge of $T + e$, and therefore there is in T a path linking the endvertices u and v of e. By substituting in D edge e by this path, we define a walk linking x and y which has one less edge which is not in T. By repeating this substitution process, as long as there is an edge which is not in T in the walk under consideration linking x and y, we obtain in the end a walk, and so a path (lemma 1.1), of T linking x and y, which ends the proof. □

Consider two more useful results (trees have many useful properties!).

PROPOSITION 2.9. *Given a spanning tree T of G and an edge e of G which does not belong to T, the spanning subgraph $T + e$ contains only one cycle.*

Proof. First of all, $T + e$ contains a cycle, according to proposition 2.8. If the edge e belonged to two distinct cycles, we could deduce in T two distinct paths linking its endvertices x and y, considering these cycles deprived of the edge e. This would contradict proposition 2.3. □

LEMMA 2.2 (Exchange lemma). *Given a spanning tree T of G, an edge e of G which does not belong to T and an edge f of the cycle of $T + e$, then $T + e - f$ is a spanning tree of G.*

Proof. In applying condition (2) of theorem 2.1, on the one hand $T + e - f$ is connected because the edge f is not a bridge of $T + e$ since it belongs to a cycle, while on the other hand we have $m_{T+e-f} = m_T = n_T - 1 = n_{T+e-f} - 1$. □

This last result gives an idea of the different spanning trees which exist in a (connected) graph, obtained by exchanging edges (Figure 2.3). It allows the generation of other spanning trees from one of them.

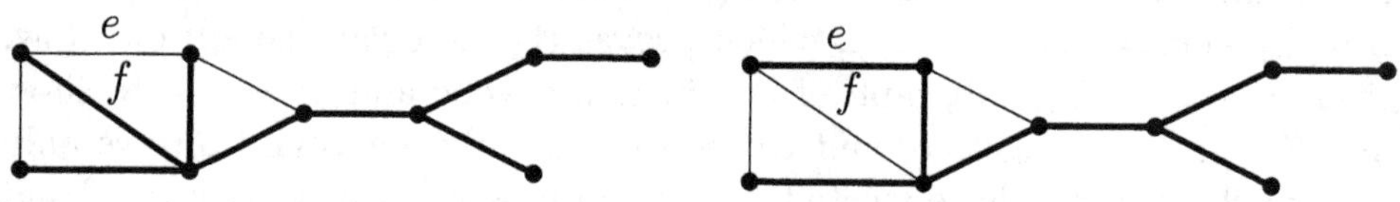

Figure 2.3. *Two spanning trees of a graph (in bold); the second one is obtained from the first one by exchanging the two edges e and f*

The following lemma, which is stronger, will be very helpful later on. The notation $T' \setminus T$ designates here the set of the edges of T' which are not in T. We can likewise define $T \setminus T'$.

LEMMA 2.3 (Strong exchange lemma). *Given two spanning trees T and T' of G, and an edge $e \in T' \setminus T$, there exists an edge $f \in T \setminus T'$ such that $T + e - f$ and $T' + f - e$ are spanning trees of G.*

Proof. By choosing edge f in the cycle of $T + e$ and not belonging to T', which is always possible since T' does not contain any cycles, we will have truly that $T+e-f$ is a spanning tree (according to lemma 2.2). Nevertheless, this will not necessarily mean that $T' + f - e$ is also a spanning tree. Edge f must be well chosen. Graph $T' - e$ has two connected components, C_1 and C_2. There is necessarily in the cycle of $T + e$ an edge, other than e, whose endvertices are one in C_1 and the other in C_2 (why?). Let this edge be f. This edge is thus not in T'. Edge e is necessarily an edge of the cycle of $T' + f$ (because e and f are the only edges of $T' + f$ joining C_1 and C_2), and so $T' + f - e$ is a spanning tree by application of lemma 2.2. Thus we have found an edge $f \in T \setminus T'$ so that $T + e - f$ and $T' + f - e$ are spanning trees of G. □

Figure 2.4 illustrates the proof. Observe in particular that of the three edges of the cycle of $T + e$ which are not in T', there is only one which is suitable (the one denoted by f in the figure).

2.2.1 An interesting illustration of trees

If we think of the theory of vector spaces, it is impossible to fail to observe the analogy of the properties of spanning trees with the bases of vector spaces. In particular, propositions 2.7 and 2.8, and especially lemma 2.2,

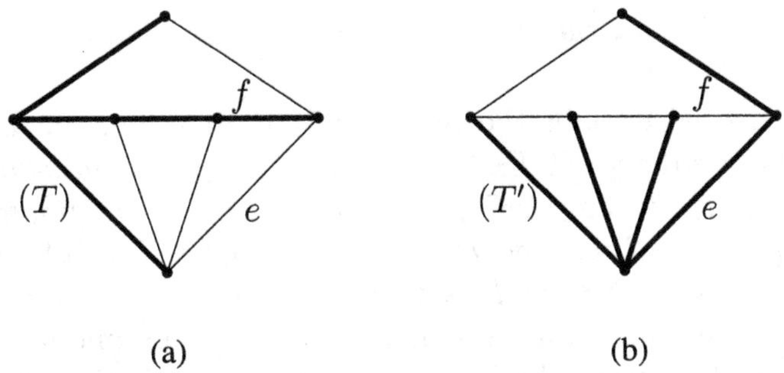

Figure 2.4. *In bold: (a) the spanning tree T, (b) the spanning tree T'*

bring to mind the classic exchange property between the bases of vector spaces. This proximity is not fortuitous and can be clarified in the following way. Let $G = (X, E)$ be a connected graph. It is possible to define a vector space on the set E in which a set of edges $F \subseteq E$ is linearly independent if, by definition, the induced spanning subgraph $G(F)$ is acyclic, and F is a spanning subset if $G(F)$ is connected. A basis of this vector space is thus a subset F which is linearly independent and which spans E, that is, such that the spanning subgraph $G(F)$ is both acyclic and connected, that is a spanning tree of G. The sets of edges of the spanning trees of G are therefore the bases of this vector space. With propositions 2.7 and 2.8 we recognize the classic characterization of the basis of a vector space: a *minimal spanning* subset or a *maximal linearly independent* subset. We find directly the finite dimension of this vector space: it is the number of edges common to all spanning trees, that is $n-1$, where n is the number of vertices of the graph.

This algebraic aspect of the graphs is the starting point of a very important and interesting theory, the theory of *matroids*.

2.2.2 Spanning trees in a weighted graph

Let $G = (X, E)$ be a graph weighted by a mapping $v : E \longrightarrow \mathbb{R}^{*+}$ (real positive numbers). Let us call $\mathcal{T}_G$ the set of the spanning trees of G. Two spanning trees T and T' are called *neighbors* in $\mathcal{T}_G$ if there are two edges e and f such that $T' = T + e - f$ and $T = T' + f - e$. For each $T \in \mathcal{T}_G$, denote $v(T)$ the sum of the values by v of the edges of T: $v(T) = \sum_{e \in A} v(e)$, where A is the set of edges of T. The elements of $\mathcal{T}_G$ are ordered by their values $v(T)$. We have in this ordered set the interesting following property.

LEMMA 2.4. *In $\mathcal{T}_G$, a local minimum is an absolute minimum.*

Proof. First let us explain the meaning of local minimum and absolute minimum of the ordered set $\mathcal{T}_G$. An element $T \in \mathcal{T}_G$ is an *absolute minimum* if for any $T' \in \mathcal{T}_G$ we have $v(T') \geq v(T)$. It is a *local minimum* if for any $T' \in \mathcal{T}_G$ *neighbor of* T, we have $v(T') \geq v(T)$. So, let T be a local minimum of $\mathcal{T}_G$ and let us show that T is also an absolute minimum. Reason by contradiction, supposing that T is not an absolute minimum. Let T_m be a spanning tree which is an absolute minimum, chosen so that the number of edges of T_m which are not in T is as small as possible. This number of edges is > 0 since $T_m \neq T$ because of the hypothesis on T. We have an edge e of T_m which is not in T and with $v(e)$ minimum. Let f be the edge of T given by application of lemma 2.3 to T, T_m and e, that is an edge f so that $T + e - f$ and $T_m + f - e$ are spanning trees of G. We have $v(e) = v(f)$. On the one hand $v(T) \leq v(T + e - f)$ because T is a local minimum, so, since $v(T + e - f) = v(T) + v(e) - v(f)$ we have $v(e) \geq v(f)$. On the other hand, $v(T_m) \leq v(T_m + f - e) = v(T_m) + v(f) - v(e)$ because T_m is an absolute minimum, thus $v(f) \geq v(e)$. Consider spanning tree $T'_m = T_m + f - e$: we have $v(T'_m) = v(T_m)$, according to the preceding equality, and thus T'_m is also an absolute minimum in $\mathcal{T}_G$. Nevertheless, T'_m has one more edge in common with T than T_m (edge f). This contradicts the hypothesis that T_m has a minimum of edges not in T. □

2.3 Application: minimum spanning tree problem

2.3.1 The problem

This problem is, in a simplified version, a communication network problem. For example, given the costs of the links between pairs of centers, we want to find a network at the lowest possible total cost. In terms of graphs, the general problem can be formulated in the following way:

`Input`: a simple connected graph $G = (X, E)$ weighted by application v with values in the strictly positive real numbers, which can represent *costs*, *distances*, *time spans*, etc.

`Output`: a spanning subgraph of G, $T = (X, A)$, where $A \subseteq E$, which is connected and such that $v(T) = \sum_{e \in A} v(e)$ is minimum.

It is easy to see at first that a solution T is necessarily a spanning tree of G. Otherwise, when applying proposition 2.7, there would be an edge e of T such that $T - e$ would always be connected. Thus, we would have $v(T - e) = v(T) - v(e) < v(T)$ (because $v(e) > 0$), which would contradict the fact that $v(T)$ was minimum. What we are then looking for is a minimum spanning tree of G. It is an absolute minimum in the ordered set T_G defined above (section 2.2.2).

2.3.2 Kruskal's algorithm

An *exhaustive* research of a solution, that is research which examines all possible cases, is generally intractable. For example, for G a graph with 20 vertices, and at a rate of examining one billion spanning trees per second (which is quite a respectable speed), it would take more than 83,125 centuries to examine all of them! The solution might be found just a little too late... This is typically an *optimization* problem. We have to find in a set of a very large number of elements, an element which is optimum, minimum or maximum, according to certain values. This is not a theoretical problem since we know *a priori* that (at least one) such element exists. It is essentially a practical problem because it is impossible on a human time scale, even with powerful computers, to find such an element by trying to look at all of them. We have to imagine a different method. For some problems, it has been possible to find a method which gives a result within a reasonable time frame, at least within certain limits. This is the case for the problem of the minimum spanning tree with the following algorithm, in which `F` is the set of edges which have been considered and `A` the set of edges selected. This classic algorithm is called *greedy* because it takes at each stage what is most advantageous, here an edge with the lowest value possible. This operating method does not lead in general to an optimal solution for optimization problems. Indeed, it often happens that a choice appearing advantageous at one moment has to be paid for later by a forced choice which is less advantageous and which takes away more than the advantage gained earlier. The problem of the minimum spanning tree is a particular case for which the greedy algorithm always works. We will see why later on. For the construction of a spanning tree, this algorithm relies on the property stated in proposition 2.8 by adding one by one, as long as possible, an edge which does not create a cycle with those already selected. In fact, this construction can be stopped as soon as the required number of edges for a tree has been reached, that is the number of vertices of the graph minus one (remember that the graph is supposed to be connected).

```
procedure Kruskal(G,v);
begin
  F:= E; A:= ∅;
  while |A| < n-1 loop
    find e ∈ F such that v(e) is minimum;
    F:= F - {e};
    if G(A ∪ {e}) acyclic then
      A:= A ∪ {e};
    end if;
  end loop;
  -- G(A) is a minimum spanning tree
end Kruskal;
```

Figure 2.5 gives an example of an application of this algorithm.

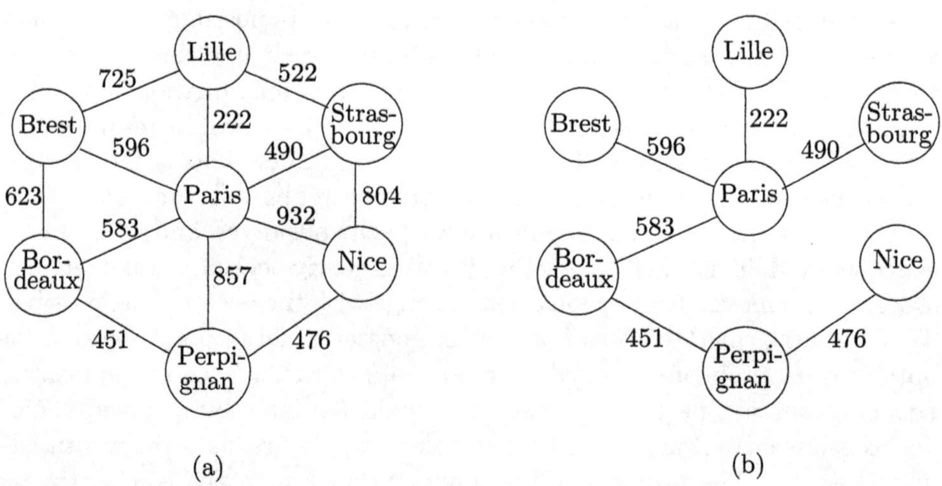

Figure 2.5. *(a)A weighted graph (representing French cities and their distances in kilometers); (b)a minimum spanning tree (communication network between these cities with the minimum total length) obtained by Kruskal's algorithm*

2.3.3 Justification

1) For any data, an algorithm must stop after a finite number of operations, that is within a finite time. The justification of an algorithm must, in particular, prove this. This is the *proof of finiteness.* This proof is easy here because the execution stops, in the worst case, when all edges of the graph have been considered. This case may happen when graph G is itself a tree. Note that here the hypothesis of the connectedness of the graph is essential: indeed we reach the exit condition of the **while** loop formula, $|A| = n - 1$, only because there really is a spanning tree in G (otherwise we would end with $F = \emptyset$ and $|A| < n-1$, making it impossible to find $e \in F$).

2) Let us show that at the end of the execution of the algorithm, spanning subgraph $T = G(A)$, induced by set A of edges, is a minimum spanning tree. First, it is a spanning tree of G by construction, since it is acyclic and verifies the relation $m = n - 1$ which comes from the exit condition of the **while** loop, $|A| = n - 1$. We still have to show that it is minimum. According to lemma 2.4, it is sufficient to show that T is a local minimum of the set $\mathcal{T}_G$ defined above (see section 2.2.2). Suppose that there is in $\mathcal{T}_G$ a neighbor $T' = T + e - f$ of T, where $e \notin T$ and $f \in T$, such that $v(T') < v(T)$. Then $v(e) < v(f)$ and when we chose edge f in the algorithm, we should have chosen e which is of lower value and which also verified the condition $G(A \cup \{e\})$ acyclic (since $T + e - f$ is acyclic). This neighbor T' of T thus cannot exist.

We can imagine the problem of the minimum spanning tree in G as a sort of "in the field" research. The points of this field are the elements of $\mathcal{T}_G$, two of them being neighbors in that field if the corresponding trees are neighbors in the sense defined above. Each of these points has a "height" proportional to the value of the tree which it represents, that is $v(T)$ for the tree T. A solution to this problem corresponds to a point of lowest height, that is a "hole" of largest depth possible in this field. Lemma 2.4 shows that this field has only holes of one depth, a hole also being without landings on its slopes. A solution is then easy to find: from any point in this field, by walking along from neighbor to neighbor toward a lower point, we end up necessarily at a solution point. That is indeed why the greedy algorithm works. In addition, we can also specify that this field has in fact only one hole (see exercise 2.19).

Other optimization problems have a more distressed "field" and their solution is therefore much less simple.

2.3.4 Implementation

In order to make this expression explicit in algorithm pseudocode and to move toward a computer program, several points must be specified and developed.

First, we must specify how the graph dealt with is supposed to be represented: the entry in the graph being done through the edges, the correct representation of G here is clearly by a list of weighted edges (see Chapter 1). Remember that in such a computer implementation of a weighted graph, we have the list of edges and for each edge its endvertices and its value.

Concerning the algorithm itself, two points need to be explained (the rest of the algorithm is direct to code):

– `find e ∈ F such that v(e) is minimum`

This operation can be done easily by sorting the edges of G in increasing order of their values. It is then sufficient to consider the edges successively in that order. It is the easy way out, but it is efficient considering the existence of good classic sorting algorithms. On the other hand, it has the practical inconvenience of spending time sorting out edges which may not even be considered later, since it is possible to have obtained a tree before having considered all edges of the graph.

– the test `G(A ∪ {e}) acyclic`

This point requires more work. What we are dealing with is the management of the connected components of graph $G(A)$ as edges are added † progressively in A. Indeed we see that $G(A \cup \{e\})$ is acyclic if and only if the endvertices of e belong to different connected components of $G(A)$. Algorithmically, an efficient way of proceeding is to regroup by lists the vertices of the connected components of $G(A)$. At the beginning, when $G(A)$ is empty of edges, each list contains one vertex. During execution, when an edge is added in A, two components, that is two lists, must be merged. In order to merge lists, it is practical to implement them as linked lists, these fusions then being done by simple assignments of pointers. It is necessary to know for each vertex the list in which this vertex lies in order to apply the preceding criteria. This last point makes it necessary, when merging two lists, to go through one of them, the one going into the other, to update the list number of its vertices (the lists being assumed to be numbered). Since we must go through one of the two lists, it is advantageous, for complexity, to choose to go through the shorter of the two lists.

2.3.5 Complexity

It is necessary to evaluate the main time-consuming actions, which are the preliminary sorting of edges and the management of the connected components of $G(A)$. With regard to sorting, a classic quick sort of m edges gives a complexity $O(m \log m)$. For the management of the connected components of $G(A)$, it is slightly more complicated. Taking into account all we have developed before, as long as we know the connected components in which each vertex is located, the test $G(A \cup \{e\})$ acyclic can be done in constant time. All it takes is to verify that the endvertices of e belong to two distinct connected components. However, the most time-consuming operation in the management of connected components is the *fusion* of two components when an edge e is added in A. With an implementation by linked lists, as described earlier, and with the choice for each fusion to include the shortest list in the largest one, as suggested above, we can obtain a fairly good complexity, $O(m \log n)$. Some more sophisticated algorithmic techniques give better results with a complexity $O(m\alpha(n))$, where α is such a slowly increasing function that its value remains practically ≤ 4 (for those with the knowledge, let us add that α is the inverse of the Ackermann function). Except for the part dealing with the sorting of edges, we thus have a complexity which is *practically linear*.

We can now remember that *if the edges of the given graph are already sorted, Kruskal's algorithm is practically linear.*

2.4 Connectivity

The graphs under consideration in this section are assumed to be simple.

In the preceding application to a communication network, we left aside a quite important practical aspect: the vulnerability of the network, that is its capacity to withstand the failure of some of its links or centers.

In terms of graphs, we consider, for example for a connected graph, the largest number of edges it is possible to remove without the graph losing its property of connectedness. With the idea of bridges we have seen this type of property earlier in this chapter. The equivalent exists for vertices.

A *cut vertex* of a graph G is a vertex x such that $G - x$ has at least one more connected component than G. This idea leads to a classic and useful decomposition of graphs.

2.4.1 Block decomposition

A *block* of a graph G is a maximal induced connected subgraph without a *cut vertex* (of itself as a graph).

† We verify the following properties of blocks (do these verifications as exercises):

– *the blocks define a partition of the set of the edges of* G;

– *two blocks may share at the most only one vertex of* G, *that vertex is then a cut vertex of* G. *Conversely, a cut vertex of* G *is a vertex shared by at least two blocks of* G.

The set of blocks of G constitutes the *block decomposition* of G. It is unique and it is a finer decomposition of the graph than the one defined by the connected components. Many properties or graph processings can be brought back to these blocks. There are some good algorithms (with linear complexity) which determine the blocks of a graph.

Figure 2.6 gives an example of block decomposition of a graph.

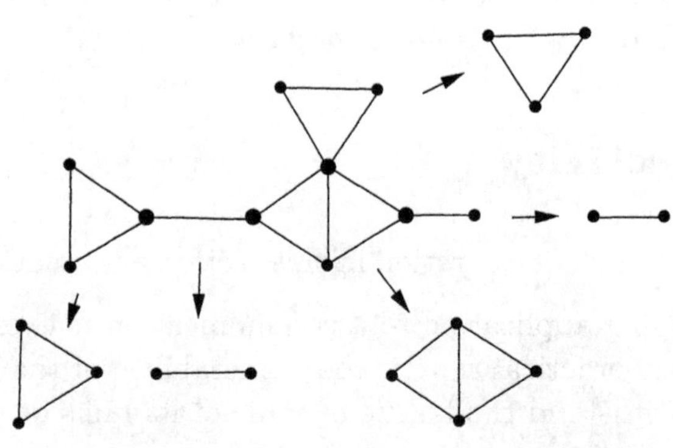

Figure 2.6. *A graph and its blocks (the cutvertices of the graph are in bold)*

2.4.2 k-connectivity

To understand the motivation of what will follow, consider the three connected graphs in Figure 2.7. The first one can be disconnected by the deletion of a vertex, x, which is a *cut vertex* of the graph.

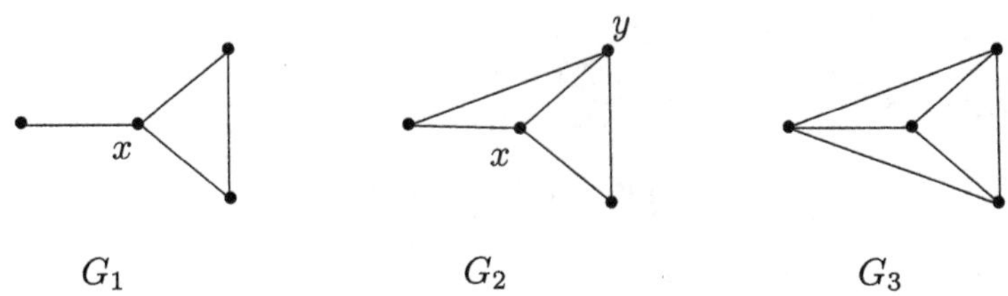

Figure 2.7. *Three examples of graphs and their connectivity:* $\kappa(G_1) = 1$, $\kappa(G_2) = 2$, $\kappa(G_3) = 3$, $\kappa'(G_1) = 1$, $\kappa'(G_2) = 2$, $\kappa'(G_3) = 3$

This is not the case with the second graph, which nevertheless can be disconnected by the deletion of two vertices, x and y. As to the third graph, it has no set of vertices by which deletion would disconnect it. In fact, this graph is complete and the only thing that can be done to it by deleting some vertices is to reduce it to a single vertex (remember that a graph has by definition at least one vertex). Looking at edges instead of vertices leads to similar observations concerning the smallest number of edges of the graph by which deletion would disconnect the graph. However, if the graph has at least two vertices, it is always possible to disconnect it by deleting some edges (we do not have the equivalent of the preceding third case for vertices). If we see these graphs as models of communication networks, we understand the importance of these considerations concerning the vulnerability to breakdowns. We introduce a parameter of a graph which measures these properties. The *connectivity* $\kappa(G)$ of a graph G is defined as the smallest number of vertices by which deletion in G yields a disconnected graph or a graph reduced to one vertex.

Let us formalize this definition. If there is in graph G a set of vertices A, which may be empty, such that $G - A$ is disconnected, then:

$$\kappa(G) = \min\big(|A| \mid G - A \text{ disconnected}\big)$$

otherwise:

$$\kappa(G) = n - 1$$

(where n is the number of vertices of G).

The case $\kappa(G) = n - 1$ is characterized by the fact that in graph G any two vertices are joined by an edge. In other words G is a complete graph (remember that G is simple). If that is the case, there is no set A of vertices such that $G - A$ is disconnected. If it is not the case, there are in G two vertices not joined by an edge, x and y, and $A = X \setminus \{x, y\}$ then has the property that $G - A$ is disconnected. Since $|A| \leq n - 2$, we can deduce the inequality $\kappa(G) \leq n - 2$.

Thus $k(G)$ is bounded by:

$$0 \leq \kappa(G) \leq n - 1$$

The case $\kappa(G) = 0$ corresponds to G disconnected or $n = 1$.

† The other following inequality, to be verified, is based on the fact that if A is the set of neighbors of a vertex, then $G - A$ is either disconnected or reduced to a single vertex. Considering a vertex of minimum degree δ_G, we deduce:

$$\kappa(G) \leq \delta_G$$

2.4.3 *k*-connected graphs

The following idea is easier to comprehend practically, unlike the connectivity of G which is not always easy to determine.

A graph G is *k-connected* if $\kappa(G) \geq k$.

We can characterize the non-trivial cases (for $k > 0$): 1-connected graphs are connected graphs such that $n \geq 2$, and 2-connected graphs are connected graphs with no *cut vertex* and such that $n \geq 3$.

NOTES. 1) If $k' \geq k$ then k'-connected implies k-connected.

2) In a graph, the blocks which have at least three vertices may be seen as 2-connected components of this graph (components in the sense of induced subgraph maximal according to the property considered).

Going back over and comparing the two ideas which we have just defined, the connectivity of G and the property of k-connectivity, we can state (integer k is assumed to be ≥ 0):

1) A graph G verifies $\kappa(G) = k$ if and only if $n \geq k+1$, $G - A$ is connected for any $A \subseteq X$ such that $|A| < k$ and there exists $A \subseteq X$ such that $|A| = k$ and $G - A$ is disconnected or reduced to one vertex.

2) A graph G is k-connected if and only if $n \geq k+1$ and $G - A$ is connected for any $A \subseteq X$ such that $|A| < k$.

2.4.4 Menger's theorem

THEOREM 2.2 (Menger, vertex statement). *A simple graph G such that $n \geq k+1$ is k-connected if and only if any two distinct vertices of G are connected by k internally vertex-disjoint paths (that is pairwise with no other common vertices than their ends).*

This is one of the major theorems in graph theory. Its proof is easy in one direction: the sufficient condition, since the existence of k vertex-disjoint paths between any two given vertices prevents the existence of fewer than k vertices by which removal would disconnect the graph. Indeed, the removal of less than k vertices in the graph cannot delete k paths linking two given vertices x and y, since these paths have no common vertices other than their ends x and y (the vertices removed are distinct from x and y). The necessary condition is less evident; a proof of this is given in Chapter 8 as an application of the theory of flows. In the meantime, as an exercise, it is interesting to try to prove it for the case $k = 2$.

2.4.5 Edge connectivity

We are going to define for edges, concepts equivalent to the one mentioned above. The *edge connectivity* $\kappa'(G)$ of a graph G, with more than one vertex, is the smallest number of edges by which removal disconnects the graph. In particular, it is 0 if the graph is disconnected. The edge connectivity is considered equal to 0 if the graph has only one vertex.

We can formalize the definition in this way. If the graph G has at least two vertices, it has a set of edges B, possibly empty, such that $G - B$ is disconnected, and we put in that case:

$$\kappa'(G) = \min\big(|B| \mid G - B \text{ disconnected}\big)$$

The set of edges B is what we call an *(edge) cut* of G. If G has only one vertex, we put:

$$\kappa'(G) = 0$$

There is an inequality relation between connectivity and edge connectivity, given in the following proposition.

PROPOSITION 2.10. *For any simple graph G, we have:*

$$\kappa(G) \leq \kappa'(G) \leq \delta_G$$

The second inequality is easy, the first can be shown directly but also results easily from Menger's theorem (see later). These inequalities may be
† strict (find an example).

2.4.6 k-edge-connected graphs

The following concept corresponds to the k-connected concept defined above.

A graph G is *k-edge-connected* if $\kappa'(G) \geq k$.

We have an "edge" version of Menger's theorem, which we equally accept:

THEOREM 2.3 (Menger, edge statement). *A simple graph G is k-edge-connected if and only if any two distinct vertices are connected by k edge-disjoint paths (that is pairwise without common edges).*

NOTE. The first inequality of proposition 2.10, $\kappa(G) \leq \kappa'(G)$, is easily deduced from both statements of Menger's theorem. Put $k = \kappa(G)$ and consider any two given vertices of G, x and y. There are k vertex-disjoint paths linking x and y according to Menger's vertex statement. Therefore there is at least the same number of edge-disjoint paths linking x and y, since the vertex-disjoint property for paths implies the edge-disjoint property. This leads to G k-edge-connected and the inequality $\kappa'(G) \geq k = \kappa(G)$, from Menger's edge statement. This inequality $\kappa(G) \leq \kappa'(G)$ is in fact natural if we observe that the removal of a vertex in a graph causes the removal of all incident edges and thus generally has a greater impact on the connectivity of the graph than the removal of a single edge.

2.4.7 Application to networks

A direct application of what has been seen above concerns the vulnerability of a communication network, that is either from the failure of the nodes (vertices) or of the links (edges). For example, the problem of minimum costs dealt with earlier can be further developed with an additional constraint of minimum resistance to failure of centers or links. Consider the following general problem:

`Input`: a simple connected graph G and an integer $k \geq 1$.

`Output`: a subgraph H of G k-connected and for which the sum of the values of the edges is minimum.

With $k = 1$, we find again, as a particular case, the problem of the minimum spanning tree, described above. For $k > 1$, the problem is known to be "difficult", to be more specific, to be **NP**-complete in the sense of the theory of complexity (when this problem is put as a decision problem, refer to Appendix B). Remember that this means that there are no known polynomial algorithmic solutions to this, and that this problem is equal in algorithmic difficulty to many unsolved problems, some of which we will encounter later.

A solution can be given in some particular cases as in the following: G is a complete graph with a value equal to 1 on each edge (see exercise 2.18).

2.4.8 Hypercube

The *k-cube*, where k is an integer ≥ 1, is a simple graph defined in the following manner: the vertices are k-tuples of 0 and 1, the edges are the pairs of k-tuples which differ by one coordinate.

Such a structure is used in parallel architecture for processor interconnection networks, in particular because of some of the following properties (see more specifically exercise 2.20):

- *The number of vertices of the k-cube is 2^k, the number of edges is $k2^{k-1}$, each vertex is of degree k (the k-cube is k-regular).*
- *The k-cube is k-connected, more specifically its connectivity number is equal to k.*
- *Any two given vertices of the k-cube are connected by a path of length $\leq k$, and this upper bound is tight.*

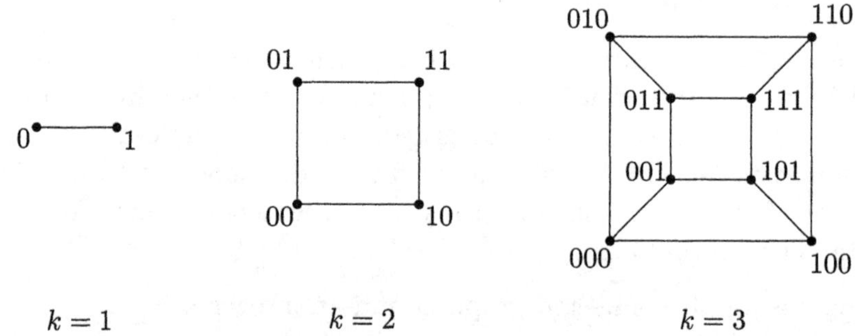

Figure 2.8. *The k-cube for* $k = 1, 2, 3$

The last property concerns what is called the *diameter* of a graph: the maximum of the distances between two vertices (the distance between two vertices being the shortest length, length meaning the number of edges, of the paths linking them). This parameter is important for a network since it corresponds to the maximum of the time required to communicate between two elements of this network, for example the transmission time of a message from one processor to another in a processor interconnection network. The k-cube has thus a diameter equal to k, a quantity which is expressed in relation to the number of vertices by $\log_2 n$ since $n = 2^k$, which means that the diameter of the k-cube does not increase too rapidly when the number of vertices increases. On the other hand, the k-cube has the inconvenience of a vertices degree which is rather high, and requires in practice a high number of input-output for each element of the network, which is not always feasible.

2.5 Exercises

2.1. Find all the trees which have six vertices (there are six).

2.2. Show that a tree which has exactly two vertices of degree one is a path.

2.3. Show that a tree has at least Δ vertices of degree one (Δ being the maximum degree).

+2.4. Show that a graph G is a tree if and only if two out the three following conditions are fulfilled:

(1) G is connected,

(2) G is acyclic,

(3) We have $m = n - 1$.

2.5. An automorphism of a simple graph $G = (X, E)$ can be defined as a bijection f from the set X to itself, such that if the vertices $x, y \in X$ are neighbors in G, then $f(x)$ and $f(y)$ are also neighbors; in other words, if $xy \in E$ then $f(x)f(y) \in E$. Show that for any automorphism f of a tree $T = (X, E)$, there exists $x \in X$ such that $f(x) = x$ or there exists $xy \in E$ such that $f(x) = y$ and $f(y) = x$ (*"fixed point property": fixed the vertex or fixed edge by isomorphism f*).

2.6. Show that the set of the subtrees of a tree T verifies the *Helly property*: if a set of subtrees of T has the property that any two of these subtrees have a common non-empty intersection, then the intersection of all the subtrees of this set is not empty (reminder: a subtree of a tree T is a subgraph of T which is a tree; here the intersection of subtrees concerns the vertices).

2.7. Show that an edge e of a connected graph G belongs to any spanning tree of G if and only if e is a bridge of G. Show that e does not belong to any spanning tree if and only if e is a loop of G.

+2.8. Let $G = (X, E)$ be a connected graph. The distance between two vertices x and y of G is the shortest length of the paths linking x and y. This distance is denoted by $d(x, y)$. We call the *center* of the graph any vertex x such that the quantity $\max_{y \in X} d(x, y)$ is the smallest possible. Show that if G is a tree then G has either one center or two centers which are then neighbors.

+2.9. In applying Kruskal's algorithm, consider the case where some of the edges have equal values and see that several solutions can then be obtained.

2.10. Does an edge of the smallest possible value in a weighted connected graph G always belong to a minimum spanning tree of G? Discuss this for a number of cases.

*2.11. *(About the implementation of Kruskal's algorithm)*

Design a management by lists of the connected components of $G(A)$. Analyze the effect on the complexity.

*2.12. *(Another justification of Kruskal's algorithm)*

a) $G = (X, E)$ a simple connected graph weighted by $v : E \to \mathbb{R}^{*+}$ (real numbers > 0). F is a spanning forest of G, U a set of connected components of F, e an edge of G linking a vertex of U and one of $X \setminus U$ and such that the value $v(e)$ is minimum among all the edges joining U and $X \setminus U$. Show that if the forest F is included, as a subgraph, in a minimum spanning tree of G, then $F + e$ also has the property of being included in a minimum spanning tree of G.

b) Apply the preceding result to justify Kruskal's algorithm. Show that after each iteration of the main loop of the algorithm, the spanning subgraph $G(A)$ is a spanning forest of G included in a minimum spanning tree of G.

*2.13. *(Jarník-Prim's algorithm)*

Another algorithm solves the problem of the minimum spanning tree. It is still the "greedy" idea that is applied, but the approach is different from that of Kruskal: it is local and therefore may be more natural.

$G = (X, E)$ a simple connected graph, weighted by $v : E \to \mathbb{R}^{*+}$ (real numbers > 0). We build a sequence $X_0 \subseteq X_1 \cdots \subseteq X$ of sets of vertices and a sequence $E_0 \subseteq E_1 \cdots \subseteq E$ of sets of edges according to the following conditions:

- $E_0 = \emptyset$, $X_0 = \{x\}$, where x is a vertex chosen arbitrarily.
- Supposing X_{i-1} and E_{i-1} already built, we look for an edge $e_i = x_i y_i$ such that $x_i \in X_{i-1}$ and $y_i \in X \setminus X_{i-1}$, with $v(e_i)$ minimum. If such an edge e_i is found, we put $X_i = X_{i-1} \cup \{y_i\}$ and $E_i = E_{i-1} \cup \{e_i\}$. If there is none, we stop.

Write this algorithm in a structured algorithmic form, justify it and analyze its complexity.

N.B. The justification of the result given by the algorithm is easy with the result of the first question of exercise 2.12 above.

*2.14. Show that a graph is planar if and only if each of its blocks is planar.

2.15. Let G be a 2-connected graph such that $n \geq 3$, x and y are two vertices of G, μ_1 is path linking x and y. Show that a path μ_2 linking x and y and vertex-disjoint of μ_1 does not always exist. (*Compare with Menger's theorem for $k = 2$; this exercise is pure logic.*)

2.16. Show, without using Menger's theorem, the following result: if a graph G is k-connected ($k \geq 1$), then for each edge e of G, $G - e$ is $(k-1)$-connected.

(*We will use this property later on to demonstrate Menger's theorem.*)

*2.17. Show that if G is simple, with a minimum degree δ such that:

$$\delta \geq \frac{n+k-2}{2}$$

then G is k-connected.

*2.18. Given integers n and k such that $0 < k < n$ and k being even, the simple graph $H_{k,n}$ is defined as follows:

- $X = \{0, 1, \ldots, n-1\}$ is the set of the vertices,
- for $i, j \in X$, ij is an edge of $H_{k,n}$ if and only if there is an integer r such that $1 \leq r \leq \frac{k}{2}$ and $j = i \pm r$, the operation $\pm$ being taken *modulo* n. For example, with $n = 8$ and $k = 4$ we have in the graph $H_{4,8}$ the edge joining the vertices 0 and 7 because $7+1$ is equal to 0 *modulo* 8.

a) Represent $H_{4,8}$.

b) Using Menger's theorem, show that $H_{4,8}$ is 4-connected and 4-edge-connected.

c) Try to generalize to $H_{k,n}$ (see Chapter 12, problem 1).

2.19. Let us go back over the definition and study of the ordered set $\mathcal{T}_G$ of the spanning trees of a weighted graph G (see section 2.2.2). Given two minimum spanning trees T_m and T'_m of G, using the strong exchange lemma, show that there is a sequence of minimum spanning trees of G, $(T_m = T_1, T_2, \ldots, T_k = T'_m)$, such that two consecutive trees are neighbors. Deduce from this that the "field" under consideration in the proof of Kruskal's algorithm (section 2.3.3) has only one "hole".

2.20. a) Show that the k-cube, for $k > 1$ may be defined as an assembly, which is to specify, of two $(k-1)$-cubes.

b) Use what you have seen above to show by induction on k that the k-cube is k-connected.

Chapter 3

Colorings

Without investigating in depth the numerous and important matters related to coloring in graphs, we will give the theoretical results necessary for solving a nice application to a timetabling problem.

3.1 Coloring problems

Historically, with the four-color theorem mentioned in Chapter 1, it is a coloring problem which is at the origin of graph theory. This concerned the problem of coloring the vertices of a planar graph. The general issue is to find the *chromatic number* of a graph, planar or not. This is the lowest number of colors needed to color the vertices so that no two adjacent vertices have the same color. There are a few known applications of this type of graph coloring. However, coloring *of edges* is also considered, and we are going to study this here because it presents an interesting application to a timetabling problem, and because it also relates to another important matter in graph theory: *matchings* (studied in Chapter 7).

3.2 Edge coloring

The graphs studied are assumed to be without loops but may have multiple edges. The "simple" hypothesis thus will mean that the graph is without multiple edges. Given a graph G and an integer k, k*-edge-coloring* of G is a mapping from the set of the edges of G to a set of k elements called

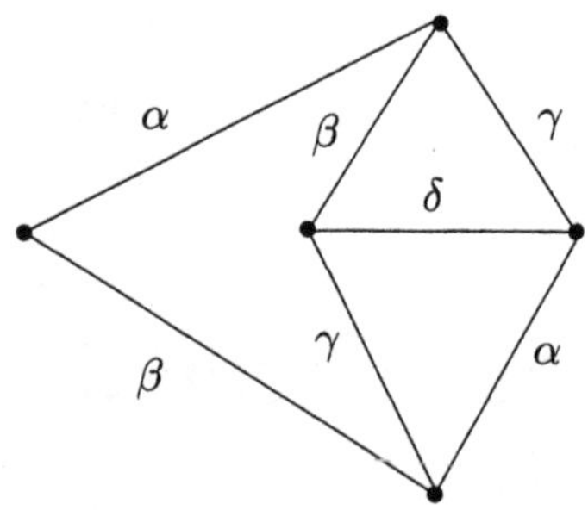

Figure 3.1. *A k-edge-coloring of a graph (set of colors: $\{\alpha, \beta, \gamma, \delta\}$)*

colors so that two edges sharing an endpoint are associated in the mapping with different colors.

Given a k-coloring, an edge is said *to be of a given color* or *to have a given color*, if in the coloring considered this color is associated with it.

The *edge chromatic number* of a graph G is the lowest integer k such that a k-coloring of G exists. This integer is denoted by $q(G)$. For example, the
† chromatic index of the graph shown in Figure 3.1 is 4 (check that for this graph there is no edge coloring with less than four colors). The important point of the concept of edge-coloring of a graph is the following property: for each color, the set of the edges having the same color forms what is called a *matching*, that is a set of edges of the graph such that no two edges share a common endpoint. A k-edge-coloring of a graph G can be seen, more or less a permutation of the colors, as an edge partition of G into matchings. The chromatic index is then the lowest number of classes of such a partition. This point of view will become useful later.

3.2.1 Basic results

It is easy to verify that the chromatic index is bounded as follows:

$$\Delta \leq q(G) \leq m$$

where m is the number of edges of the graph and Δ its maximum degree. If
† it is easy to see that the lower bound is often reached (find some examples), the upper bound m is on the other hand often too large. The following remarkable result shows that in the case of simple graphs $\Delta + 1$ is a much better upper bound.

THEOREM 3.1 (Vizing). *If G is a simple graph, then $q(G)$ is equal to Δ or $\Delta + 1$.*

The case of bipartite graphs is even more remarkable and has the following basic result, for which the proof is given as an exercise at the end of this chapter.

THEOREM 3.2. *If G is bipartite, simple or not, then $q(G) = \Delta$.*

This second result will be useful later on in solving our timetabling problem.

3.3 Algorithmic aspects

The problem of the chromatic index of a simple graph is a remarkable example of an **NP**-complete problem (even when the **NP**-completeness has been found much later than for other graph problems). Indeed, deciding if the chromatic index of a simple graph is Δ or $\Delta + 1$ is an **NP**-complete problem.

Nevertheless, the problem of finding in a simple graph a maximum matching, that is one having the largest number of edges possible, can be solved by a polynomial algorithm. Indeed, it is with this problem that the concept of polynomial complexity was introduced. It is then possible to think of the following algorithm to find the chromatic index of a graph, this index being seen here as the lowest number of classes of a partition of edges into matchings: consider a maximum matching C_1 of G, and then a maximum matching C_2 of $G - C_1$, the graph obtained by removing the edges of C_1, and so on until the removal of all edges of G. The matchings successively obtained, $C_1, C_2, \ldots$ of G, construct an edge partition into matchings of the graph. We have tried to minimize the number of these matchings by considering each time a matching with a maximum number of edges in what was left. This way of proceeding means giving a first color as often as possible to some edges of the graph, while respecting the condition of never giving the same color to two edges sharing the same endvertex. We then give a second color to a maximum number of edges not yet colored, and so forth until all edges of the graph are colored. Unfortunately, this *greedy* algorithm (see Chapter 2) in general does not yield an optimal result, that is, a coloring with the lowest possible number of colors, as shown in Figure 3.2.

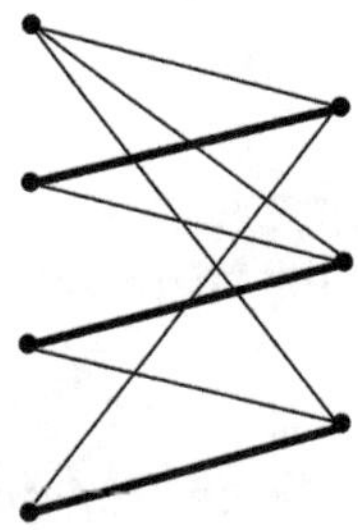

Figure 3.2. *By giving a first color to the edges of the maximum matching shown in bold, it will not be possible to have a Δ-coloring of this bipartite graph of maximum degree 3 (because of the degree 3 vertex on the upper left). Nevertheless such a coloring exists (according to theorem 3.2)*

It is always possible to try every way of giving colors to the edges to find a coloring. There is a general classic method called *backtracking*. It equates to a *depth–first search* of an arborescence (an algorithm studied in Chapter 5).
† In fact it is a very good anticipation exercise to try to implement this search for the chromatic index of the graph, called the *Petersen graph*, shown in Figure 3.3. This graph does not have a Δ-coloring; its chromatic index is not 3 but 4 (as was already the case in Figure 3.1).

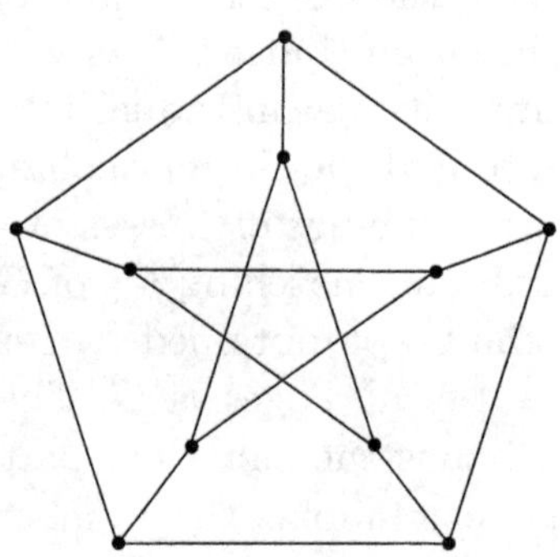

Figure 3.3. *Petersen's graph (chromatic index* = 4*)*

In fact, there is no other real general method known at present to verify that there is no Δ-coloring. However, the number of cases increases exponentially with the size of the graph, showing all too well the difficulty of an **NP**-complete problem.

The case of bipartite graphs is algorithmically more accessible. It is possible to find with a polynomial algorithm a Δ-coloring of any bipartite graph (simple or not). This is particularly interesting in relation to the timetabling problem. The study of such an algorithm is proposed as an exercise at the end of this chapter.

3.4 The timetabling problem

The simplest version of this problem (more complex ones will be studied later) is as follows: some professors have to teach courses to classes every week. These courses are defined by the number of hours given by each professor to each class. The week is assumed to be divided into time slots in which the courses have to be timetabled. The goal is to establish a weekly timetable taking into account that: a professor cannot teach two classes at the same time and two classes are not taught by two professors at the same time (that would be terrible, guess for whom ...).

The modeling of this problem in graph terms is easy: $G = (X, Y, E)$ is a bipartite graph where X is the set of the professors, Y the set of the classes and E the set of the courses, where an edge joining $x \in X$ and $y \in Y$ corresponds to a one-hour course by professor x to class y. We observe that, under the constraints of this problem, the courses located in a given time slot make up a matching in the graph (according to the definition given above). Thus, a timetable corresponds to an edge partition into matchings of the graph, a partition which can be assimilated in an edge-coloring of the graph.

First, we will address a particular question: *what is the lowest number of hours necessary to establish a timetable for all the courses to be given*? In the associated graph G this equates to finding the lowest number of matchings in which to partition the edges of the graph. This shows how the timetabling problem can be brought back to an edge-coloring problem in a bipartite graph: each time slot corresponds to a color and a timetable corresponds to a coloring. As a consequence, the answer to the question of the lowest number of hours is clear: the lowest number of hours in which it is possible to establish a timetable is simply the chromatic index of the graph G associated with the problem. Graph G being bipartite, the lowest number of hours is equal to the maximum degree Δ of G, according to theorem 3.2. A timetable with this lowest number of hours is defined by a Δ-coloring of G. It is thus very interesting to obtain such a coloring in polynomial time. Interpreted using the data of the problem, Δ is the maximum of the maximum hours due by a teacher and of the maximum course hours for a class.

3.4.1 Room constraints

In real life there is hardly any timetabling problem which is not constrained by room availability. So, in order to be more realistic, we will now introduce such a constraint. Specifically, we will suppose that the number of rooms available is r (with $r \geq 1$). Let m be the total number of course hours to be given, i.e. $m = |E|$. Note k the lowest number of hours in which it is possible to establish a timetable. We can first observe, through † overall calculation, that $k \geq \lceil \frac{m}{r} \rceil$ (specify why). With no room availability constraint, we saw that $k = \Delta$, but with this constraint, we have *a priori* only $k \geq \Delta$. Taking all this into account, we have:[1]

$$k \geq \max\left(\Delta, \left\lceil \frac{m}{r} \right\rceil\right)$$

In fact, we have an equality, as the following lemma will show.

LEMMA 3.1. *Let $G = (X, Y, E)$ be a bipartite graph and k an integer $\geq \Delta$. There is a partition of E into k matchings with the same number of elements for one unit.*

For colorings, such a partition corresponds to a coloring in which each color appears the same number of times on the edges, for one unit.

Proof. The proof can be given in a constructive way, which is important here for its application to the timetabling problem. Go from any k-coloring of G. It is always possible to obtain algorithmically such coloring with a Δ-coloring (see section 3.3) considered as a k-coloring of which $k - \Delta$ colors are unused when $k > \Delta$. The next step consists of balancing the colors among themselves. Reason in the equivalent language of matchings in considering the coloring as a matchings family $(C_1, C_2, \ldots, C_k)$. Consider a matching C_i of lowest cardinality and a matching C_j of highest cardinality, and suppose that $|C_j| > |C_i| + 1$. Spanning subgraph $G(C_i \cup C_j)$ induced by $C_i \cup C_j$ (disjoint union since C_i and C_j are disjoint), is constituted of connected

[1] For a number x, we denote $\lceil x \rceil$ the least integer $\geq x$ and $\lfloor x \rfloor$ the greatest integer $\leq x$.

components, each being an isolated vertex, a path, or a cycle. These last two types of connected components alternate in C_i and C_j, that is have their edges alternately in C_i and C_j. This implies that the cycles are necessarily even and have the same number of edges belonging respectively to C_i and C_j. The global difference of the number of elements between C_i and C_j thus can only be found in a component which is a path, to be more specific in an alternated path with respect to C_i and C_j starting and finishing with an edge of C_j, matching with the most edges. An exchange of the edges of C_i and C_j along such a path, without any other modifications, yields two new matchings C_i' and C_j' verifying:

$$0 \leq \left||C_j'| - |C_i'|\right| < \left||C_j| - |C_i|\right|$$

Indeed, the number of elements of C_i increases by 1 in C_i', while it decreases by 1 in C_j' with respect to C_j. Continuing this balancing process between the two matchings under consideration, we obtain in the end an equal cardinality to plus or minus one unit. The repetition of this process to all matchings, as long as there are two which differ by more than one cardinal unit, leads to the result given in the lemma. □

NOTE. The k matchings resulting from the lemma each have a number of elements which can only be $\lfloor \frac{m}{k} \rfloor$ or $\lceil \frac{m}{k} \rceil$, these two quantities being equal when k divides m.

This lemma allows us to state the following proposition:

PROPOSITION 3.1. *Given an instance of the timetabling problem with a number of rooms equal to r, the lowest possible number of hours for which a timetable is feasible is:*

$$\max\left(\Delta, \left\lceil \frac{m}{r} \right\rceil\right)$$

where m is the total number of hours to be given and Δ the maximum of the maximum hours due by a professor and of the maximum course hours for a class.

Proof. Apply lemma 3.1 to bipartite graph G associated with this timetabling problem, with $k = \max(\Delta, \lceil \frac{m}{r} \rceil)$ (note that we effectively have $k \geq \Delta$). According to the preceding comment, each of the k matchings given by the lemma has a number of edges less than or equal to $\lceil \frac{m}{k} \rceil$. Since

$\lceil \frac{m}{r} \rceil \leq k$, by definition of integer k, we also have $\lceil \frac{m}{k} \rceil \leq r$. Thus, each matching of this partition has a number of elements less than or equal to r. As a result, in the timetable where the courses given in time slots are defined by the preceding matchings, in each time slot the number of courses taught simultaneously never exceeds the number r of rooms. This is the constraint we wanted to satisfy. □

It is interesting to see how the room constraint is satisfied with lemma 3.1: logically, but not trivially, by spreading *as uniformly as possible* the courses over the time slots. This is the right way to minimize the number of rooms used.

NOTES. 1) It is in fact possible to obtain a timetable respecting the room availability constraint in k hours for any integer k greater than or equal to $\max(\Delta, \lceil \frac{m}{r} \rceil)$. This may be useful in order to take into account additional constraints.

2) It is possible to use inversely the formula giving the lowest number of hours according to the number of rooms, that is by wondering what is the minimum number of rooms which is necessary to build a timetable in a given number of hours (look for the lowest r so that the maximum expression is less than or equal to this number of hours).

3.4.2 An example

Suppose the following data, keeping the general preceding notations: four teachers x_1, x_2, x_3, x_4, five classes y_1, y_2, y_3, y_4, y_5, and the courses to be given defined in Table 3.1 below.

	y_1	y_2	y_3	y_4	y_5
x_1	1	2	0	0	0
x_2	1	1	1	0	0
x_3	0	1	1	1	1
x_4	0	0	0	1	2

Table 3.1. *Data*

Always in keeping with the general notation, we have here: $m = 13$, $\Delta = 4$. Thus, without a room availability constraint, a timetable in four hours is possible. This would mean at least $\lceil \frac{m}{4} \rceil = 4$ rooms available, as shown by an overall calculation (as calculated above with the number of rooms r). However, suppose there are only three rooms available, that is $r = 3$. The general result above then sets the lowest number of hours for a timetable equal to:

$$\max\left(\Delta, \left\lceil \frac{m}{r} \right\rceil\right) = \max(4,5) = 5$$

Therefore there is a timetable in five hours with three rooms (but not in four hours). In order to build such a timetable we will apply the general preceding process which can be followed in Figure 3.4. We are building a bipartite graph G associated with this problem. We then define, directly here because it is a simple case, a partition in $\Delta = 4$ matchings, C_1, C_2, C_3, C_4. We consider this partition as being in 5 matchings, the last one C_5 having no edges at present. The cardinalities of these matchings are respectively: 4 for C_1, 4 for C_2, 3 for C_3, 2 for C_1, and 0 for C_5. We will then balance three times: twice between C_1 and C_5, and once between C_2 and C_4 (note that an exchange can be made on a single edge when the alternating path is reduced to one edge as is the case for the first two exchanges). In the end we obtain an edge partition of G into matchings with the same number of elements to one unit, that is 2 and 3. Associating matchings and hours, this partition defines a timetable in five hours with three rooms: C_1 to the first hour, C_2 to the second, and so forth (see Table 3.2: the teachers are in rows, the hours in columns). Observe that there are never more than three courses in parallel during the same time slot, which shows that the room availability constraint is indeed satisfied.

Hours	1	2	3	4	5
x_1	—	—	y_2	y_2	y_1
x_2	—	y_3	y_1	—	y_2
x_3	y_3	y_2	y_5	y_4	—
x_4	y_4	y_5	—	y_5	—

Table 3.2. *Timetable given for professors (for instance, professor x_1 teaches class y_2 in hour 3)*

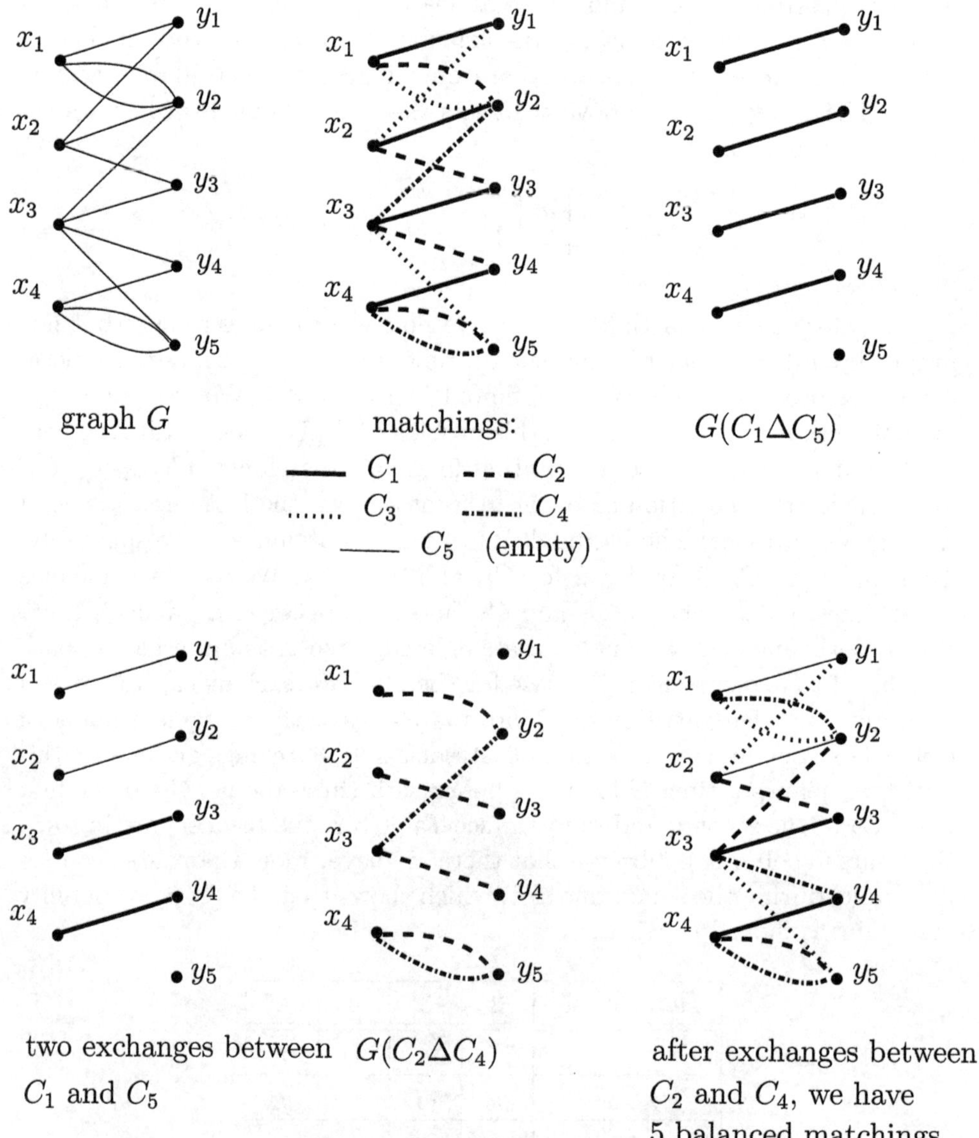

Figure 3.4. *Solving the example*

NOTE. The association of matchings and time slots is arbitrary *a priori*. It is possible to play on that factor to take into account additional constraints, for example regrouping the teaching load of x_2 in three consecutive hours by exchanging hours 4 and 5.

3.4.3 Conclusion

Real time timetabling problems are much more complicated because of the many diverse constraints which may not be so easy to formalize as the one above. For example, the rooms may not all be ordinary, some may be labs or sport rooms which are not exchangeable. There may also be constraints over certain time slots for the classes or the teachers. However, the greatest difficulty arises when classes are split and regrouped differently for options, for example. Nevertheless the preceding theory is a good initial frame of reference to start formalizing and implementing solutions. Note that on a theoretical level, the timetabling problem with constraints on the times of classes and the loads of teachers becomes **NP**-complete.

3.5 Exercises

3.1. Show that if a simple and connected graph G has an odd number of vertices n, then $q(G) = \Delta + 1$ (the solution is very simple). Apply this result to complete graph K_n.

3.2. Consider the following timetable in six hours, with four professors $p_1, \ldots, p_6$ and five classes $c_1, \ldots, c_5$.

Hours	1	2	3	4	5	6
p_1	c_1	c_3	c_5	—	c_4	c_5
p_2	c_2	c_1	c_1	c_3	—	—
p_3	—	c_4	—	c_2	—	c_1
p_4	c_4	c_5	c_2	c_1	c_3	c_3

a) How many rooms are necessary for this timetable?

b) Show, without redoing this timetable for the time being, that it is possible to establish another timetable for the same number of classes, still in six hours but with one room less.

c) Build a timetable answering the preceding question, with the additional constraint that professor p_2 must be free during the last two hours (5 and 6).

d) With the same number of classes, how many hours would be necessary to build a timetable with a room less than in question c? (You are not asked to build the timetable.)

N.B. The small size of the problem makes it possible to find solutions by hand through trial and error without using the methods described above. Such a method cannot be generalized and is therefore of no interest.

*3.3. *(Proof of $q(G) = \Delta$ for a bipartite graph)*

We will consider *pseudocolorings*, that is, simple mappings from an edge set of a graph to a set of colors (two incident edges to the same vertex may have the same color). Let G be a bipartite graph and C a Δ-pseudocoloring of the edges of G (one pseudocoloring is easy to obtain, for example by associating the same color with all the edges). Note $\alpha, \beta, \ldots$ the Δ colors. Consider the case where a vertex x of G has at least two incident edges of the same color, say α, in C.

a) Show that there is a color, say β, which appears on none of the incident edges of x.

b) We will try to "ameliorate" the pseudocoloring C by having color β appear on an incident edge of x, in place of color α. Examine the consequences of an exchange of colors α and β on one of the edges e with color α incident to x. Observe, in particular, that then α may disappear from the set of the incident colors of the other endpoint y of e. We are then led to another exchange at vertex y to bring back α. Show that this exchange process of α and β can only fail for an odd cycle with edges alternatively of color α and β, except in vertex x where both incident edges are of color α and these edges are the only two incident edges of x with that color. Conclude.

c) Show that by repeating the preceding process often enough it is possible to turn the initial pseudocoloring into a true coloration, that is such that two incident edges of a vertex are of different colors, hence the equality $q(G) = \Delta$. *(This method, using recolorings, leads to a polynomial algorithm.*[2]*)*

[2]A general method is given in the following reference: *Méthode et théorème général de coloration des arêtes d'un multigraphe*, J.C. Fournier, Journal de Mathématiques pures et appliquées, Vol. 56 (1977), 437–453.

Chapter 4

Directed Graphs

Many applications are modeled by graphs with edges which are oriented, the endvertices of the edge not playing a symmetric role. They are called directed graphs. Among them, two classes are particularly important: directed graphs without circuits and arborescences, the latter being commonly called "trees" in, for example, computer science.

4.1 Definitions and basic concepts

A *directed graph*, abbreviated to *digraph*, G, is defined by two finite sets: a non-empty set X of *vertices* and a set A of *arcs*, or *directed edges*, with an ordered pair (x, y) of vertices which are the *endvertices* of a, associated with each arc. Vertex y is called the *head* and vertex x is called the *tail*.

4.1.1 Notation

We write $G = (X, A)$. Sets X and A may also be denoted by $X(G)$ and $A(G)$. We keep, as in the undirected case, the notations n_G or n for the number of vertices and m_G or m for the number of arcs.

4.1.2 Terminology

When (x, y) is the ordered pair of endvertices associated with arc a, we say that arc a *joins* vertex x to vertex y, that arc a is *incident* to vertices x and y, or more specifically that arc a *comes out of* vertex x and that it *enters into* vertex y. Vertex y is called a *successor* of x and vertex x is called a *predecessor* of y. As with graphs, we have a *loop* in the case of equality $x = y$, and a *multiple arc* in the cases of arcs having an identical ordered pair (x, y) associated with them. It is possible to specify, according to the number of arcs involved: *double arc*, *triple arc*, etc. Two arcs of which the associated ordered pairs are respectively (x, y) and (y, x) are said to be *opposed*.

A digraph is said to be *strict* if it has no loops and no multiple arcs (it may have opposed arcs). In this case, often encountered, each arc is identified by the ordered pair of its endvertices, which are then distinct, and we write, for example, $a = (x, y)$. For a strict digraph $G = (X, A)$, we can define the set of the arcs A directly as a subset of the Cartesian product $X \times X$.

A strict digraph is *symmetric* if for any arc (x, y) there is also the opposed arc (y, x). This concept is very close to that of a graph.

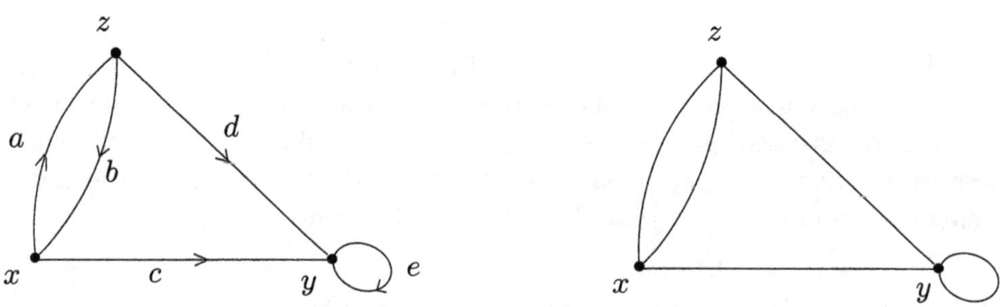

$X = \{x, y, z\}$, $A = \{a, b, c, d, e\}$

(a) (b)

Figure 4.1. *(a) A digraph: the arc a is associated with the ordered pair (x, z), b with the ordered pair (z, x), c with the ordered pair (x, y), d with the ordered pair (z, y), and e with the ordered pair (y, y); (b) the underlying graph*

4.1.3 Representation

As we saw in Figure 4.1(a), digraphs are drawn in a plane as graphs, with simply an arrow indicating its orientation on each arc line, the arrow going from x to y if the ordered pair (x, y) is associated with it.

Finally, to complete the definition of digraphs, it should be noted that the digraphs under consideration are *unlabeled* digraphs, that is, considered up to isomorphism. We will not go back over this concept which can be defined with graphs and does not pose any practical problem.

4.1.4 Underlying graph

Given a digraph G we consider the *underlying graph*, obviously defined by "forgetting" the orientation of the arcs of G, that is that each arc with which the ordered pair (x, y) is associated is replaced by an edge of which the associated pair of endvertices is xy (see Figure 4.1(b)).

NOTE. Given a graph G there are 2^m digraphs for which the underlying graph is G (m being the number of edges of G).

All concepts defined for graphs also apply to digraphs by means of the underlying graph: *degree* of a vertex, *connectedness* and *connected components*, *walk*, *trail*, *path*, etc. For example, we say that a digraph is connected if its underlying graph is connected. These concepts are therefore independent from the orientation of the arcs. There are also other concepts which take into account the orientation of the arcs and which we are now going to define.

4.1.5 "Directed" concepts

Some concepts can be transposed directly from the undirected to the directed case by replacing the word *edge* by the word *arc*. These are: *subdigraphs*, *induced subdigraphs*, *spanning subdigraphs*, *directed walks*, *directed trails*, *directed paths* and *directed cycles*, which we also call *circuits*. For example, a *directed walk* is a sequence of elements, alternately vertices and arcs beginning and ending with a vertex, and such that each arc has for its tail the preceding vertex in the sequence and for its head the following vertex. The first and the last vertices are the *ends*. A *directed cycle*, or *circuit*, is a closed directed path of length ≥ 1 (its ends coincide). We say that a directed walk or a directed cycle *goes through* an arc or a vertex if it contains this arc or vertex. The *length* of a directed walk or a directed cycle is the number of its arcs. As with graphs, a directed walk in a digraph may have zero length but a cycle must have a length ≥ 1.

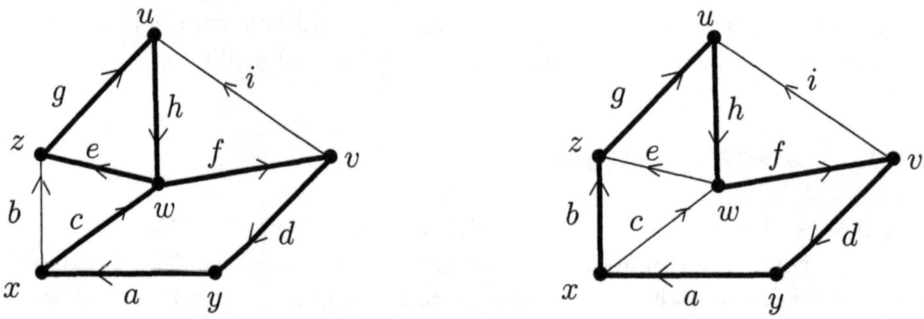

(a) (b)

Figure 4.2. *A strict directed graph, with, in bold: (a) a directed closed trail* $(x, c, w, e, z, g, u, h, w, f, v, d, y, a, x)$*; (b) a circuit* $(x, b, z, g, u, h, w, f, v, d, y, a, x)$

The following concepts have a definition specific to the directed case.

4.1.6 Indegrees and outdegrees

The *indegree* of a vertex x of a digraph G is the number of arcs entering into x. The *outdegree* of x is the number of arcs exiting from x. These are denoted by $d_G^-(x)$, or $d^-(x)$, and $d_G^+(x)$, or $d^+(x)$, and are integers.

† It is easily seen that:

$$\sum_{x \in X} d_G^-(x) = \sum_{x \in X} d_G^+(x) = m$$

Loops are not often considered in digraphs, nevertheless this formula correctly accounts for them, knowing that each loop in a digraph counts for one unit of outdegree and one unit of indegree of the vertex under consideration. We again find the formula for the sum of the degrees given in Chapter 1, in observing that for each vertex x we have: $d_G(x) = d_G^-(x) + d_G^+(x)$, so:

$$\sum_{x \in X} d_G(x) = \sum_{x \in X} \left(d_G^-(x) + d_G^+(x)\right) = \sum_{x \in X} d_G^-(x) + \sum_{x \in X} d_G^+(x) = 2m$$

4.1.7 Strongly connected components

A digraph G is strongly connected if for any two distinct vertices x and y there is a directed path going from x to y. Note that as this definition is symmetric in x and y, there is also a directed path from y to x.

A *strongly connected component* of G is a maximal strongly connected induced subdigraph of G. Maximal means that there is no strongly connected induced subdigraph containing strictly (for the vertices) this subdigraph. These components may also be defined as being the subdigraphs induced by the classes of the following equivalence relation on the vertices: there is a directed path from x to y and a directed path from y to x. These components define a partition of the set of the vertices (but not of the arcs) and constitute *decomposition into strongly connected components* of G. This decomposition is unique.

The strongly connected components are less simple to determine than the connected components, but there are nevertheless good algorithms (of linear complexity) for finding them.

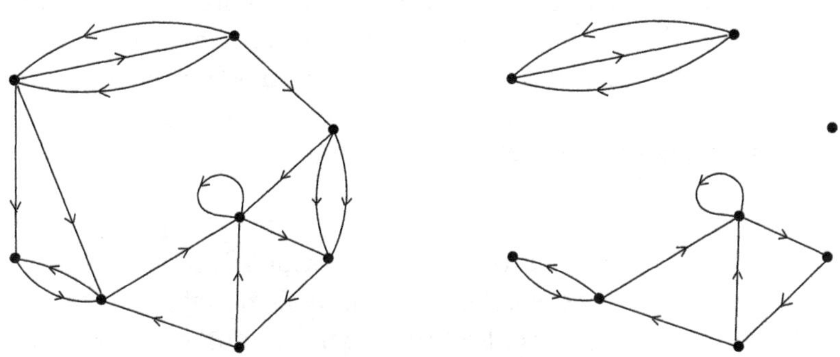

(a) (b)

Figure 4.3. *(a) A digraph which is not strongly connected; (b) its three strongly connected components (note that one of them is reduced to a vertex)*

NOTE. Vertices belonging to the same circuit belong to the same strongly connected component (justify). †

4.1.8 Representations of digraphs inside a machine

The same principle used to represent a (undirected) graph inside a machine can be applied to digraphs, with a few additional specifications because of orientation.

1. *The adjacency matrix* $M = (m_{ij})$ of a digraph $G = (X, A)$, with $X = \{x_1, \ldots, x_n\}$, is defined by putting m_{ij} equal to the number of arcs of which the associated ordered pair is (x_i, x_j). Contrary to the undirected case, this square matrix is not usually symmetric (but it is the case with a symmetric digraph).

2. The data for the neighborhood of each vertex can be given by *lists of successors*: each list associated with a vertex contains the successors of that vertex. Classically, it is possible to implement these lists as linked lists. In concrete terms, a table indexed on the vertex type contains for each vertex an access pointer to its list of successors. For some applications (for example for the potential task graph in Chapter 6) the digraph is more naturally represented by *lists of predecessors*. Each vertex is associated with the list of its predecessors. It is not very hard to go from one to the other, that is from successors' lists to † predecessors' lists and vice versa (write the corresponding algorithms). Another way to model a digraph using the neighborhoods of its vertices (which we will use in Chapter 8 for flows) is to give for each vertex the set of arcs exiting from this vertex and the set of arcs entering this vertex.

3. The *list of arcs* constitutes the third principle of the implementation of digraphs. In concrete terms, it is also possible to have an array indexed on the type arc, the arcs being numbered from 1 to m, with the ordered pair of its endvertices associated with each arc. These data can be given in a record composed of two fields, in the order: the tail of the arc, then its head.

The advantages and inconveniences of these various representations, in particular the memory space required, can be analyzed as in the case of graphs.

Figure 4.4 illustrates these representation principles.

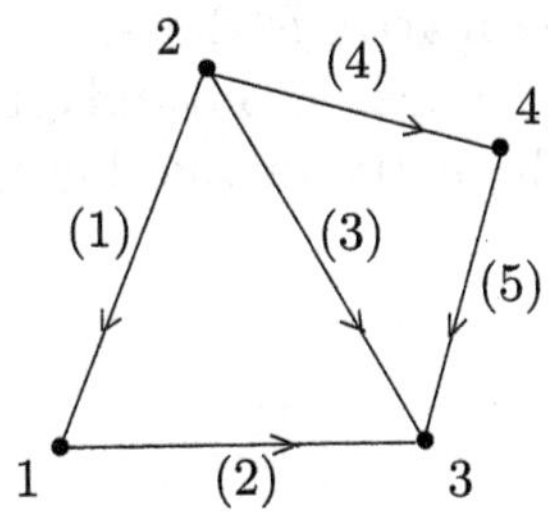

digraph

	1	2	3	4
1	0	0	1	0
2	1	0	1	1
3	0	0	0	0
4	0	0	1	0

adjacency matrix

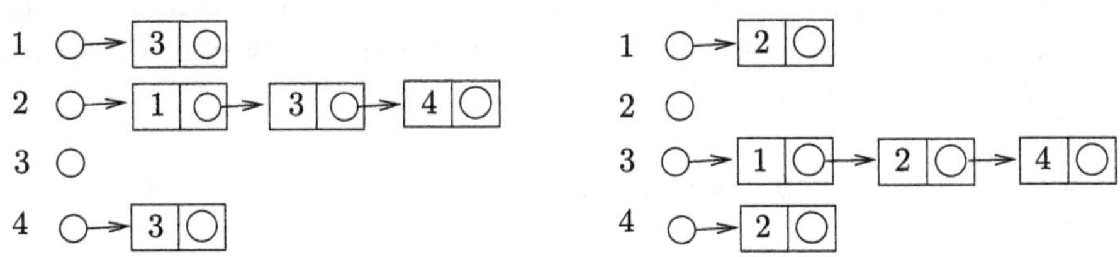

lists of successors

lists of predecessors

arcs	ordered pairs of endvertices
1	(2,1)
2	(1,3)
3	(2,3)
4	(2,4)
5	(4,3)

list of arcs (array)

Figure 4.4. *Different representations of a (strict) digraph*

We also consider *weighted* digraphs: $G = (X, A)$ with a mapping $v : A \to \mathbb{R}$. The implementing of these digraphs by an adjacency matrix, when they are strict, is generally well suited: each entry of the matrix is the value of the corresponding arc. The implementing by list of arcs can also be adapted but in practice the choice is made depending on the entry in the digraph required by the algorithm.

4.2 Acyclic digraphs

Digraphs without circuits, or *acyclic* digraphs, have important applications, notably in scheduling with the potential task graph (defined in Chapter 6). We will give below one very useful characteristic property of these digraphs.

In general, a *source* of a digraph is a vertex with a zero indegree, that is a vertex without entering arcs. Likewise, a *sink* is a vertex with a zero outdegree, that is a vertex without exiting arcs.

LEMMA 4.1. *In a digraph without circuits, there are a source and a sink.*

Proof. Consider any vertex x_1, then, if it exists, a predecessor x_2 of x_1 and a predecessor x_3 of x_2, and so on as long as a predecessor to the vertex under consideration can be found. This construction of a sequence of vertices necessarily stops after a finite number of vertices. Indeed, the digraph being finite and having by hypothesis no circuit, it is not possible to encounter a vertex seen previously again. When this construction stops, we have a vertex which by construction has no predecessor, that is we obtain a source of the digraph, the existence of which is thus proven. It is possible to proceed likewise for a sink (or to consider the *converse*, that is, the digraph obtained by reversing the direction of the arcs: a source vertex of one is a sink vertex of the other). □

4.2.1 Acyclic numbering

An *acyclic numbering* of the vertices of a digraph $G = (X, A)$ is a bijection f from X onto the interval of integers from 1 to n (n is the number of vertices of G), such that if the ordered pair (x, y) is associated with an arc of G, then $f(x) < f(y)$. The digraph is considered to be connected (underlying graph) and strict.

4.2.2 Characterization

PROPOSITION 4.1. *A digraph G is without circuits if and only if it allows an acyclic numbering of its vertices.*

Proof. The sufficient condition is easy to verify. Indeed, if there were a circuit defined by the sequence of its vertices $(x_0, x_1, \ldots, x_0)$, we would have $f(x_0) < f(x_1) < \cdots < f(x_0)$ and therefore a contradiction since then $f(x_0) < f(x_0)$. For the necessary condition we can consider a source vertex x_1 of G (which we know exists from lemma 4.1) and put $f(x_1) = 1$, then continue with digraph $G_1 = G - x_1$ in place of G, that is consider a source vertex x_2 of G_1 for which we put $f(x_2) = 2$. We remove x_2 from G_1 and start again with $G_2 = G_1 - x_2$ in place of G_1, and so on, while there is at least one vertex left in the current digraph. Note that each of the graphs successively considered, $G_1, G_2, \ldots$, is really without circuits, as it is a subdigraph of G. It is easy to verify that, by construction, the application f thus defined is an acyclic numbering of the vertices of G. □

NOTE. The algorithmic aspect of the construction of acyclic numbering is in practice very important. The preceding proof does give an algorithm but one which is not very good from an implementation and complexity viewpoint. Indeed, the removal of the vertex in the digraph complicates things. We will see a much better algorithm later, using a depth-first search of the digraph (see Chapter 5).

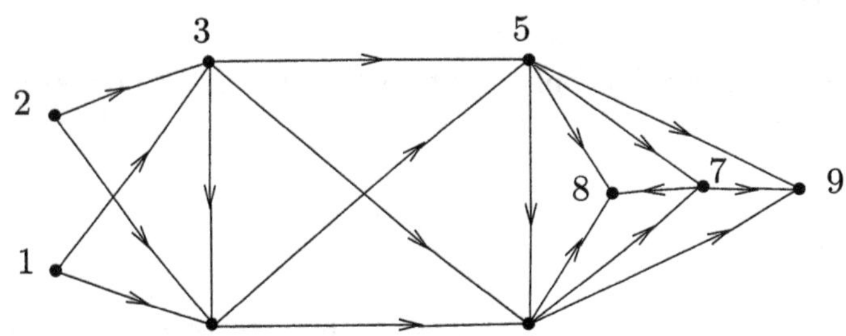

Figure 4.5. *A digraph without circuits with its vertices numbered according to an acyclic numbering. The vertices 1 and 2 are source, the vertices 8 and 9 are sink*

4.2.3 Practical aspects

In practice, we use a concept close to acyclic numbering of vertices, which is a classification of the vertices by *levels*. The important property is then that at each level a vertex has only predecessors at lower levels (see Chapter 6, exercise 6.6). We have a similar property to acyclic numbering, but the acyclic numbering is more directly usable since we always end up looking at the vertices in a certain order. Let us therefore remember what is most essential here, which is how we will apply it all later on: *when we consider the vertices of a digraph in the order of an acyclic numbering, at the time when a given vertex is considered, all its predecessors have been considered.* Thus, if we have to apply a certain treatment on each vertex of a digraph without circuits, and this treatment uses the result of this same treatment on the predecessors of the vertex under consideration, then this treatment done in the order of an acyclic numbering will be possible on all the vertices of the digraph.

4.3 Arborescences

The *root* of a digraph G is a vertex r such that there is for any vertex x of G a directed path from r to x.

An *arborescence* is a digraph which has a root and of which the underlying graph is a tree. In the literature, the term *tree* is often used instead of *arborescence*.

NOTE. An arborescence has only one root (but generally speaking, a digraph may have several roots).

4.3.1 Drawings

We usually draw an arborescence with the root at the top (which would make us say that here the trees have their roots up in the air!), and the other vertices arranged horizontally by levels of equal distances from the root. The arrows indicating the orientation of the arcs are occasionally omitted as they are implicitly defined as going from top to bottom (see Figure 4.6).

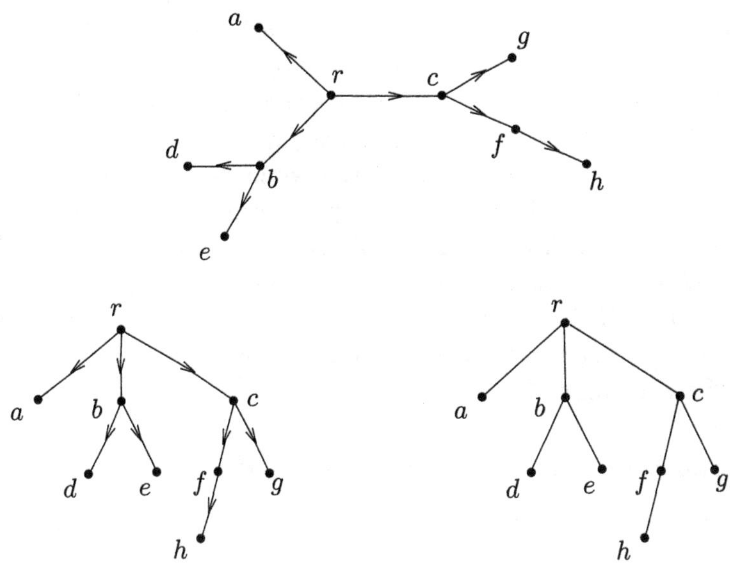

Figure 4.6. *Different drawings of an arborescence*

NOTE. An arborescence is identified by a (undirected) tree with a distinguished vertex called the root. The orientation of the arcs may be found by following the direction of the increasing distance from this root on each edge. (Note that this direction is uniquely defined because of the uniqueness of the directed path from the root to a vertex in an arborescence, according to a property given later on.) That is why arborescences are often called *rooted trees* in the literature, or simply *trees*.

4.3.2 Terminology

An arborescence, as for any digraph without circuits, has at least one sink and usually there are many. Such a vertex is called a *leaf*. In general the vertices of an arborescence are called *nodes*.

The *depth* of a vertex is its distance from the root ("distance" here means the shortest length of the directed paths). This term refers to the usual drawing of arborescences, in which, as we saw, the vertices with the same depth are placed on what is called a same depth *level*.

The *depth* of an arborescence is the greatest depth of its vertex.

Terminology inspired by that of genealogical trees can also be used. A *child* of a vertex is any successor to that vertex. A vertex without a child is a leaf. If vertex y is a child of vertex x, x is then naturally called the *parent* of y. Any vertex of an arborescence has a unique parent except for the root, which has none. Vertices which share the same parent are then also very naturally called *siblings*. This terminology – parent, child, sibling – is useful for "surfing" in an arborescence as we will see in the following chapter. More generally, a *descendant* of a vertex x of an arborescence is the name given to any vertex y such that there is in the arborescence a directed path going from x to y. In other words, y is a successor or, in a repeated fashion, a successor of a successor of x.

4.3.3 Characterization of arborescences

THEOREM 4.1. *For a digraph G, the following conditions are equivalent:*

(1) G is an arborescence.

(2) G has a root and its underlying graph is acyclic.

(3) G has a root and $m = n - 1$.

(4) There is a vertex r such that for any vertex x of G there is a unique directed path going from r to x.

(5) G is connected and $d^-(x) = 1$ for any vertex x except for one vertex, r, for which $d^-(r) = 0$. (This last condition will be referred to below as the "indegrees condition".)

(6) The underlying graph of G is acyclic and we have the property of the indegrees condition.

(7) G has no circuits and we have the indegrees condition.

Proof (sketch). Let us give a "complete" set of implications, complete in the sense that any other implication may be deduced from it by transitivity (we apply here a concept of "transitive closure", see exercise 4.3).

† The implications to verify, chosen for their ease, are:

– equivalence of conditions (1), (2), and (3);

– equivalence of (6) and (7);

– (1)⇒(4), (4)⇒(5), (5)⇒(6), (6)⇒(1). □

NOTE. In condition (4), the unique directed path from vertex r to itself is the directed path with of zero length and unique vertex r.

4.3.4 Subarborescences

Given an arborescence T and one of its vertices s, the *subarborescence* of root s is the subdigraph of T induced by s and all its descendants in T. This subdigraph is an arborescence, with vertex s for its root.

4.3.5 Ordered arborescences

It is frequent in applications to consider what can be called an *ordered arborescence*, that is an arborescence in which an (total) order is given to the set of the children of each vertex. In general, this is the case in practice since an arborescence is often defined for each vertex by the data of its children in a list, that is with a certain order. In the case of an ordered arborescence, it is possible to speak of the *first child* or of the *following sibling* of a vertex (as we will do, for example, in the following chapter).

Looking at the arborescence in Figure 4.6, and more specifically the bottom drawings representing an ordered arborescence (from left to right for the children of each vertex), we see for example that: d is the first child of b, e is the following sibling of d, e has no following sibling.

4.3.6 Directed forests

A *directed forest* is a digraph in which each connected component is an arborescence. The underlying graph is of course a forest, and a directed forest can be identified with a forest in which each connected component has a distinct vertex which is its root. The properties of directed forests can be deduced from those of the arborescences applied to its components.

4.4 Exercises

+4.1. Go over exercise 1.4 in Chapter 1, adapting it to digraphs.

+4.2. Go over exercise 1.11 in Chapter 1, adapting it to digraphs.

4.3. The *transitive closure* of a strict digraph $G = (X, A)$ is the digraph $\overline{G} = (X, \overline{A})$ such that $(x, y) \in \overline{A}$ if and only if $x \neq y$ and there is a directed path from x to y in G. Show that G is strongly connected if and only if G is the symmetric complete digraph, that is such that for any two distinct vertices x and y, (x, y) is an arc.

*4.4. Show that a digraph G is strongly connected if and only if it is connected and each of its blocks is strongly connected.

*4.5. *(Problem of one way streets in a town)*

Let G be a connected graph. Show that there is an orientation of G which gives a strongly connected digraph if and only if G is 2-edge-connected.

4.6. Show that a digraph is without circuits if and only if all its directed walks are directed *paths.*

*4.7. Show that a digraph G has no circuit if and only if its adjacency matrix M is *nilpotent*, that is, such that there is an integer $k \geq 1$ for which $M^k = 0$.

4.8. The *reduced digraph* of a digraph G is the strict digraph in which the vertices are the strongly connected components $C_1, \ldots, C_p$ of G and the arcs are the ordered pairs (C_i, C_j) such that there is an arc from a vertex of C_i to a vertex of C_j. Show that the reduced digraph is without circuits.

4.9. Show that a digraph is an arborescence if and only if it has a root and is minimal for this property concerning the removal of arcs.

Chapter 5

Search Algorithms

From a general point of view, a *search* of a graph, or a digraph, is an algorithm which makes it possible to search the arcs of the graph and to visit its vertices with a special purpose in mind. This chapter presents one of the most classic of these searches, called a *depth-first search* (often abbreviated *dfs*). This type of tree-search will be completed in Chapter 6 by another classic tree-search, the *breadth*-first search.

In applications which are modeled by an arborescence, this search technique is called *backtracking*, and can be used to solve a wide variety of problems in operations research and artificial intelligence.

5.1 Depth-first search of an arborescence

From an algorithmic viewpoint, the *recursive* form is the most natural and most efficient to express this search. The following procedure expresses the depth-first search of the subarborescence of arborescence T, of root v. We designate as `children(v)` the set of the children of vertex parameter v.

```
procedure dfs_arbo_recu(T,v);
begin
  for u in children(v) loop
    dfs_arbo_recu(T,u);   -- recursive call
  end loop;
end dfs_arbo_recu;
```

The complete arborescence is searched by a call of this recursive procedure on the root r of the arborescence T, that is, the call: `dfs_arbo_recu(T,r)`.

In practice, as in the applications we will see later, an arborescence is usually given by what may be called "navigation primitives", meaning subprograms which allow a child or a sibling of a given vertex to be reached. In order to be more specific, the children of the vertex are usually given in a particular order defined by a list. Remember that it is then an *ordered* arborescence, and that we can refer to the *first child* of a vertex when this vertex is not a leaf, and to the *following sibling* of a vertex if it has one. We express the preceding search with the following primitives in which the names indicate what they are for: **`exists_sibling(v)`** is a Boolean function which returns *true* or *false* depending on whether the vertex parameter `v` has or has not a child in the arborescence; **`first_child(v)`** returns the first sibling, when it exists. We similarly define the primitive **`exists_following_sibling(v)`**, which returns *true* or *false* depending on whether or not there is a following sibling of the vertex parameter `v`, and **`following_sibling(v)`** which returns the first following sibling (the one just following), when it exists.

```
procedure dfs_arbo_recu(T,v);
begin
  if exists_child(v) then
    u:= first_child(v);
    dfs_arbo_recu(T,u);   -- recursive call
    while exists_following_sibling(t) loop
      u:= following_sibling(u);
      dfs_arbo_recu(T,u);   -- recursive call
    end loop;
  end if;
end dfs_arbo_recu;
```

5.1.1 Iterative form

It is interesting to eliminate the recursion in the preceding algorithm to follow the search strategy step by step. One of the possible iterative forms is given below. We could have used a "parent" primitive of the arborescence, which would have made it possible to avoid having to use a stack. In fact, in practice it is easier to do without this primitive by using a stack.

In addition, this stack recalls the stack used by the computer system for the management of recursive calls and recursive returns during execution. The stack primitives used here are classic. Let us specify that **pop(S)** removes the element which is at the top of stack S, *without returning it*, and `top_stack(S)` returns the element which is at the top of stack S, without removing it from S. Variable vertex v is local. It represents the current vertex of the search. Parameter r represents the root of arborescence T, also passed as a parameter of the procedure. Remember that the instruction **exit** causes exit from the current loop.

```
procedure dfs_arbo_ite(T,r);
begin
  push(S,r);
  v:= r;
  loop
    while exits_child(v) loop
      v:= first_child(v);
      push(S,v);
    end loop;
    -- exits_child(v) false
    while v ≠ r and then not exists_following_sibling(v) loop
      pop(S);
      v:= top_stack(S);
    end loop;
    -- v = r or exists_following_sibling(v) true
    exit when v = r;
    pop(S);
    v:= following_sibling(v);
    push(S,v);
  end loop;
  pop(S);
end dfs_arbo_ite;
```

NOTES. 1) As formulated, with the **and then** (the second condition is tested only if the first is verified), the **exit** condition of the second **while** loop ensures that primitive `exists_following_sibling` will not be called on root r, which makes it possible to avoid having to plan this particular case for this primitive. This case is in fact useless since the root of an arborescence never has a sibling.

2) Stack `S` is initially assumed to be empty. It is then also empty at the end of the execution. Indeed, the last pop, after the main **loop**, exits vertex `r`, which is the first and last vertex in the stack.

The strategy of this search may be described in natural language by following the moves of the current vertex v (moves defined by the successive assignment of the variable v in the algorithm). Initially, v is in r. Then at each step of the search the current vertex v goes: to the first child of v if v has a child, to the following sibling of v if v has no child or v no longer has any child left unconsidered but has a following sibling, and finally, v goes to the parent of v if v no longer has any child or following sibling left unconsidered but has a parent, that is if $v \neq r$. The search ends when v is back to r. This description reveals the priority for vertex v to move first to a child, what can be called the "depth-first descent", and explains the source of the terminology "depth-first search". The given algorithmic expression shows this strategy through the layout of the loops. The first interior loop (**while**) corresponds to the onward search, the second interior loop (**while**) corresponds to returning to the parent. The exterior loop (**loop**) corresponds to a move toward the following sibling, between the onward and upward searches expressed by the two preceding loops.

5.1.2 Visits to the vertices

As we will see, the use of the search is made through some timely appropriate actions while visiting the various vertices of the arborescence. It is possible to specify these visits in terms of *previsits* or *postvisits*, and equally to spot the visits through the leaves of the arborescence, which are often important. This is done as commentaries of the following version of procedure `dfs_arbo_ite`, which completes the iterative version given earlier. These visits are easy to spot in the recursive version (procedure `dfs_arbo_recu`), because *each previsit corresponds to a push* and *each postvisit corresponds to a pop*. So, there is a *previsit* of the current vertex before a recursive call on this vertex and there is a *postvisit* at the time of the recursive return. This corresponds to the iterative version since, in the management of the recursion, pushes correspond to recursive calls and pops to recursive returns (except for the root).

```
procedure dfs_arbo_ite(T,r);
begin
  -- previsit of r
  push(P,r);
  v:= r;
  loop
    while exists_child(v) loop
      v:= first_child(v);
      -- previsit of v
      push(S,v);
    end loop;
    -- v is a leaf
    while v ≠ r and then not exists_following_sibling(v) loop
      pop(S);
      -- postvisit of v
      v:= top_stack(S);
    end loop;
    -- v = r or exists_following_sibling(v) true
    exit when v = r;
    pop(S);
    -- postvisit of v
    s:= following_sibling(v);
    -- previsit of v
    push(S,v);
   end loop;
   pop(S);
   -- postvisit of v
 end dfs_arbo_ite;
```

The numbering defined by the previsits is classically called the *preorder numbering* of the vertices of the arborescence, and the numbering defined by the postvisits the *postorder numbering*. Figure 5.1 gives an example of a depth-first search of an arborescence, with preorder and postorder numbering of the vertices.

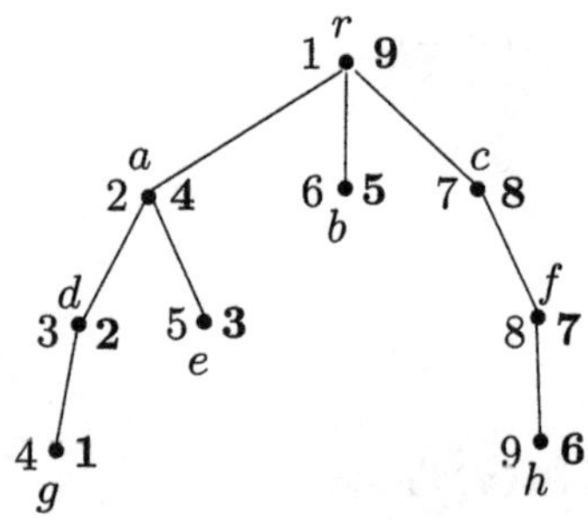

Figure 5.1. *To the left of each vertex is given its preorder number and to the right, in bold type, its postorder number*

5.1.3 Justification

It is easy to be convinced that this algorithm, under any of the versions presented, really performs a search of the arborescence in the sense given above. Indeed, through systematic consideration for each vertex of all its children, each arc is considered and each vertex is visited.

5.1.4 Complexity

Time complexity for this search, as a function of the number of vertices of the arborescence, is linear. Effectively, for each vertex its children are considered only once as children of this vertex. The total number of elementary operations is thus of the order of the sum of the outdegrees of the vertices, that is of the order of the number, m, of arcs of the arborescence. Since $m = n - 1$, where n is the number of vertices, the complexity is $O(n)$. However, if we measure the size of the arborescence by its depth d (the greatest length of a path from the root to a leaf), rather than by the number of vertices, which is much more relevant in applications, the search complexity becomes exponential. For example, the complexity becomes $O(k^d)$ for an arborescence for which any non-leaf vertex has k children and any leaf is of depth d. This complexity is in fact proportional to the number of vertices visited. We will see later the concrete consequences of this exponential complexity.

5.2 Optimization of a sequence of decisions

Let us consider the problem of having to choose one decision among several possibilities at each step of a process. Some states of the process, called *terminal*, no longer require a decision and can be assessed with a *gain* that is an integer or real number, which may be positive, negative or zero. The problem, then, is to find a decision sequence which leads from a given *initial* state to a terminal state with the greatest possible gain.

Modeling this problem by an arborescence is easy: each vertex represents a state, the root represents the initial state, each arc corresponds to a possible decision leading from one state to another. The leaves are the final states and they are assigned to values corresponding to the gains. We have to determine in this arborescence a path from the root to a leaf which has the greatest possible gain value.

As formulated, and with what has been developed above, there is a direct solution to this problem: a depth-first search of the arborescence, recording the values of leaves as they are visited, should bring about a solution. In practice, things are less simple. On the one hand, searching a whole arborescence may be costly in time and frequently even impossible to complete within a reasonable human time scale. On the other hand, the arborescence associated with the problem is not fully known from the start and has to be built, which is not crippling but requires some technical work. Let us illustrate this with a classic algorithmic problem.

5.2.1 The eight queens problem

This is an old puzzle, already known in the 19th century. The aim is to put eight queens on a chessboard so that none of them is able to take another, according to chess rules. Recall that the queen attacks any piece that is in the same row, column or diagonal. It is clear that it is impossible to place more than eight queens. The question therefore is how to find out if it is possible with eight queens.

It is necessary to define the arborescence associated with the problem because several possibilities are conceivable. Since there must not be more than one queen per row, one natural way to proceed is to put one queen per row, row after row, starting from row one (supposing that the rows are numbered from 1 to 8, for example from bottom to top). The root of the

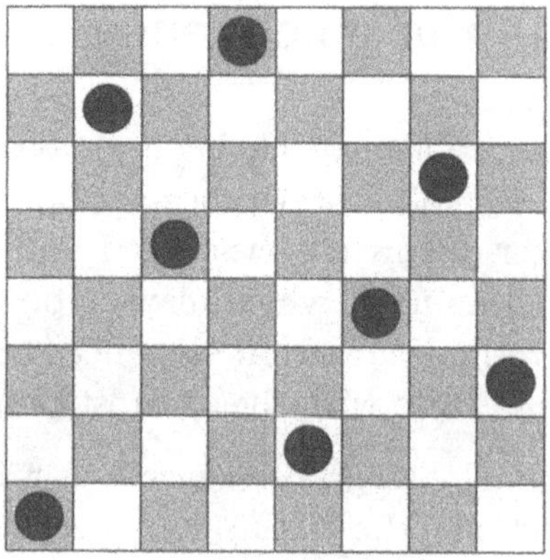

Figure 5.2. *A solution to the eight queens problem*

arborescence is the empty chessboard. The children of the root correspond to the eight ways to place a queen on the first row. The children of the children correspond to the ways of placing a queen on the second row out of reach of the preceding queen, and so on for all the following rows. A leaf of the arborescence corresponds to a state of the chessboard where some queens are already placed on a number of first rows in such a way that it is impossible to add another one on the following row, either because there is in fact no following row or because all the squares of the following row are under the threat of the previously placed queens. The gain associated with an arborescence leaf is the number of queens placed on the board, a number which corresponds to the depth of the leaf in the arborescence. A solution is reached when eight queens are placed, that is for a leaf of depth 8, which is the greatest possible gain.

For the search application to this arborescence, the real work is in defining the arborescence primitives, `exits_child`, `first_child`, `exists_following_sibling`, `following_sibling`. For example, the function `exits_child` must return *true* or *false*, depending on whether there is or is not in the row following the current one a free square where a queen can be placed, taking into account the ones already placed. The function `first_child`, in the case where `exits_child` has returned *true* for the vertex under consideration, must return the first following free square, deciding, for example, to go from left to right on the squares of each row. The complete algorithmic resolution of this problem is proposed as an exercise at the end of this chapter.

5.2.2 Application to game theory: finding a winning strategy

The games under consideration here are *two-player games*, with *complete information* (each player has complete knowledge of the entire game), *hazardless* (no throwing dice or drawing cards for example) and *with zero sum* (the sum of the gains of the two players is zero). Chess is a typical example. The player who starts is denoted by A and the other one by B. Let us consider only simple gains, that is: 1 if A wins (then B loses), -1 if A loses (and B wins), 0 if the game is tied. For this type of game, there is always what is called a *winning strategy*, that is a way of playing for one of the players which, regardless of the moves of the opponent, ensures that he or she does not lose. This means A has a gain ≥ 0 (if A has a winning strategy), B has a gain ≤ 0 (if B has a winning strategy). It is possible to prove that there is always a winning strategy, but it is not possible to say *a priori* which of the players has it. It depends on the game, and both cases really happen.

5.2.3 Associated arborescence

We are going to show how a winning strategy can be found algorithmically, which will constitute a constructive proof of its existence. To do this, let us associate an arborescence with the game in order to apply a search to it. Its root is of course the initial state of the game. Its children are all the states of games obtained after the first move of player A, the children of the children all the states obtained after the move then made by B, and so on, alternating moves by A and B. The leaves are states of the game obtained by a sequence of moves, which then represent a match, and after which there are no more moves, the game being over. For each leaf, the gain is assessed as defined above: 1 if A wins, -1 if B wins and 0 for a tie.

NOTE. A given state of the game may appear several times as a vertex of the arborescence. This is inherent to this model of the game, since a given situation in a game can generally be obtained by different sequences of moves.

5.2.4 Example

The arborescence of a game is soon enormous (the preceding remarks contribute to that effect)! To present a case which remains accessible, we are going to consider the game of Nim. In its classic form (popularized by a French film in the 1960s), there are four piles of matches containing respectively one, three, five, and seven matches. In turn, each player removes as many matches (but at least one) as he or she wishes, from only one pile. The player removing the last match has lost. We will limit ourselves to a reduced version of this game with only two piles of one and three matches. Its arborescence is fully represented in Figure 5.3. Each state of the game is coded by the data of the number of matches in each pile separated by a hyphen, 1-3 for the initial state for example. Note that despite the great simplicity of this game, its arborescence is already a bit complicated.

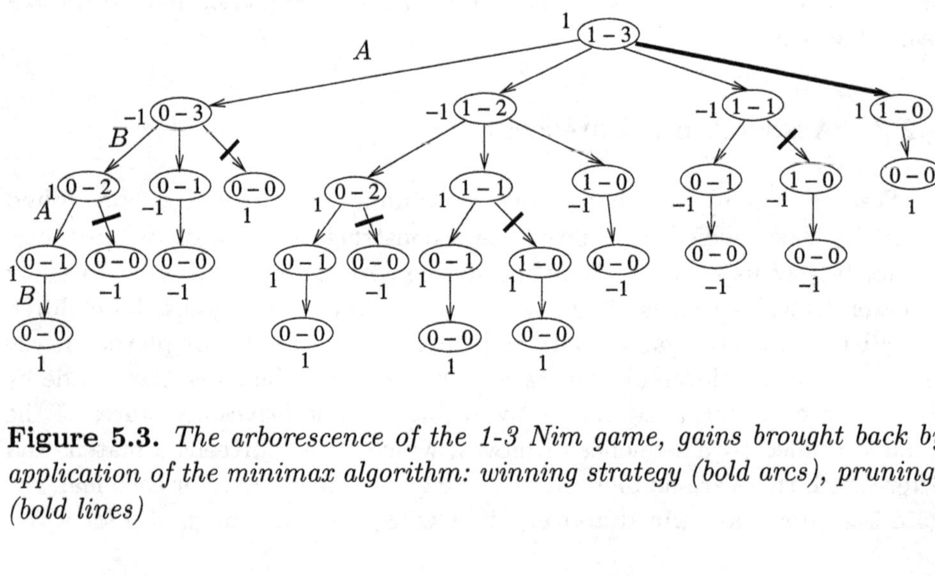

Figure 5.3. *The arborescence of the 1-3 Nim game, gains brought back by application of the minimax algorithm: winning strategy (bold arcs), prunings (bold lines)*

5.2.5 The minimax algorithm

Finding a winning strategy is harder than the simple optimization of a sequence of decisions as described above. Indeed, there is an antagonistic pursuit between the players: the one who starts is looking for a maximum gain, since it is directly his gain. The other is trying to minimize the final gain since his/her gain is the opposite. The essential principle of the algorithm is that of the *mounting of the gains of the leaves*, values which are known, towards the root, where the returned value will indicate which of the players is benefiting from a winning strategy. In general, the gain returned for a vertex of the arborescence will equal 1 if it is a *winning position* for player A, -1 if it is a *losing position* for A (and therefore winning for B), 0 if it is not a winning position for either of the players. This last situation is that of a tied game, possible for each of the two players if no "mistake" is made, that is the case when each player avoids, as is possible, making a move leading to a winning situation for his opponent. In this particular case, we can say that each player has a winning strategy in the sense defined above (a strategy which would be better called a "non-losing" strategy).

The principle described indicates by itself how to return the gains. Let us take for example the case of a state of the game where it is A's turn, and let us suppose that the gains of all the children of this vertex in the arborescence have already been determined. Then the rule to apply is that the gain returned to the vertex under consideration is the maximum of the gains of its children. That will also define what is the "best move" for A to make at this moment. A similar formula applies in the case where it is B's turn to play, with *minimum* instead of *maximum*. Thus, from bottom to top, that is, from leaves to the root, it is possible to find the values of the gains returned for each vertex of the arborescence and finally for the root. This technique of alternately returning a minimum and a maximum explains the name *minimax* given to this algorithm.

5.2.6 Implementation

A depth-first search of the arborescence of a game is perfectly adapted to the implementation of this technique of gain return. At each postvisit of a vertex the value of its parent is updated by *max* or *min* depending on whether it is a move by A or B which is returned. Indeed, at this point all the children of the vertex under consideration have known gain values.

In order to avoid singling out the postvisit case of the first child of a vertex, a point at which the parent does not yet have a returned value, from the case of the following children, that is to be able to apply one single formula, we initialize in previsit the value of each vertex in the following manner: -1 if it is A's turn, +1 if it is B's turn. These values are chosen in such a way that the value returned the first time will be taken automatically. For example, at the first return of the child of a vertex representing a game situation where A is about to play, the maximum will be taken between -1 and a gain g returned equal to -1, +1 or 0. Therefore, the taken gain will necessarily be g. The same result can be obtained for B with a minimum between +1 and a gain equal to -1, +1 or 0.

5.2.7 In concrete terms

To effectively find a winning strategy, it is necessary to keep a record of the arc which gave the best return gain so far for each vertex throughout the search. Figure 5.3 illustrates this method. In concrete terms, the application of this minimax algorithm to a realistic game is impossible to do within a reasonable time period, so enormous is the arborescence to be searched. For example, it is impossible to search the entire chess game arborescence (the number of leaves of its arborescence can be evaluated to be 20^{50}). Even if it was possible, the storage of the information for a winning strategy would create problems. Such a strategy must give the right answers for all the possible moves of the opponent. We are facing the "combinatorial explosion" phenomenon, in this case the exponential growth of cases to contemplate.

5.2.8 Pruning

In order to calculate the gain returned to the root, it is possible to reduce the search by *pruning*, that is to "prune" the arborescence. This technique does not modify the exponential nature of the search. Figure 5.3 shows pruning cases which can be understood on their own. For example, when value 1 is returned to vertex 0-2, which is at depth 2 at the bottom left, from vertex 0-1, it is useless to explore the other branch leading to vertex 0-0 since the returned value 1 is the best possible for player A, whose turn it is in this situation. Indeed, whatever the value returned from vertex 0-0, it will not modify the value previously returned in vertex 0-1.

Even though the minimax algorithm does not allow a global exhaustive search, it is nevertheless useful for a *local* search, that is a search which does not necessarily continue until the end of the game but limits itself, for example, to a 10-move exploration depth from the situation analyzed. Finding the best move possible is then done on the basis of evaluation functions which quantitatively assess game situations at the limit of the exploration, situations which are not yet end game and thus without known gains *a priori*. This technique is used by chess game software; their high level of performance is well known and is due essentially to the quality of these evaluation functions which contain in fact all the human "expertise" in this matter.

5.3 Depth-first search of a digraph

The digraph G searched here is assumed to be strict (there are no parallel arcs, no loops, and any arc is identified by the ordered pair of its ends). The digraph is also supposed to be given by *linked* lists of successors, and in the following algorithmic expressions `suc(v)` designates a pointer to an element of the list of vertex `v`, initially to the first element of this list. Each of these elements is a record[1] which contains: `suc(v).vertex` the next successor of `v` to be examined, `suc(v).next` a pointer to the following element of the list, equal to **null** if there are no more elements in it. When a successor has been read, `suc(v)` must be incremented, that is moved to point to the following element of the list of successors. This is what is produced by the assignment `suc(s):= suc(s).next`. The array `visited`, indexed on the vertices of the digraph, is supposed initialized to the value *false* for each vertex.

In a recursive form, we first have the following procedure, with parameter `G` as the digraph to be searched and `v` as the initial vertex of the search:

```
procedure dfs_recu(G,v);
begin
  visited(v):= true;
  while suc(v) ≠ null loop
    u:= suc(v).vertex; suc(v):= suc(v).next;
    if not visited(u) then
      -- previsit of u
      dfs_recu(G,u); -- recursive call
      -- postvisit of u
    else
      -- revisit of u
      null;
    end if;
  end loop;
end dfs_recu;
```

The main procedure for a search starting at a given vertex `r` of `G`, which is the vertex origin of the search and which is passed as a parameter, is written as follows:

[1] *record* in Ada, *struct* in C.

```
procedure dfs(G,r);
begin
  -- previsit of r
  dfs_recu(G,r);
  -- postvisit of r
end dfs;
```

5.3.1 Comments

We have first given the recursive form, which is more concise. It is useful to locate all the different types of visits at the vertices in this algorithm, because it will be of the greatest use in the application of this search. The *previsits* and the *postvisits*, which correspond, as for the above arborescence, respectively to *recursive calls* and *recursive returns*, are mentioned as comments. The case of the *revisit* of a vertex is new, compared with the case of the arborescence. It corresponds to the case of a vertex which has already been visited and is encountered again as successor to the current vertex (a case which may not happen for an arborescence since there is only one path from one vertex to another). This case appears in the **else** of the **if** and since there is then nothing to do, this is made explicit by the instruction **null**.[2]

The following second form of the depth search is iterative and corresponds to recursion elimination of the preceding one. It is therefore the same search strategy. It is instructive to follow the evolution of the stack in this search, in particular for *previsits* and *postvisits* which, as in the arborescence case, correspond respectively to *pushes* and *pops*. The array `visited` is still assumed to be initialized to the value *false* for each vertex of the digraph.

```
procedure dfs_ite(G,r);
begin
  -- previsit of r
  visited(r):= true;
  push(S,r); v:= r;
  loop
    while suc(v) ≠ null loop
```

[2]Be sure to distinguish the *statement* **null**, as here, and the *value* **null** of a pointer which points to nothing, as seen above.

```
      u:= suc(v).vertex; suc(v):= suc(v).next;
      if not visited(u) then
        -- previsit of u
        visited(u):= true;
        push(S,u); v:= u;
      else
        -- revisit of u
        null;
      end if;
    end loop;
    pop(S);
    exit when empty_stack(S);
    -- postvisit of v
    v:= top(S);
  end loop;
  -- postvisit of r
end dfs_ite;
```

Figure 5.4 gives an example of an application.

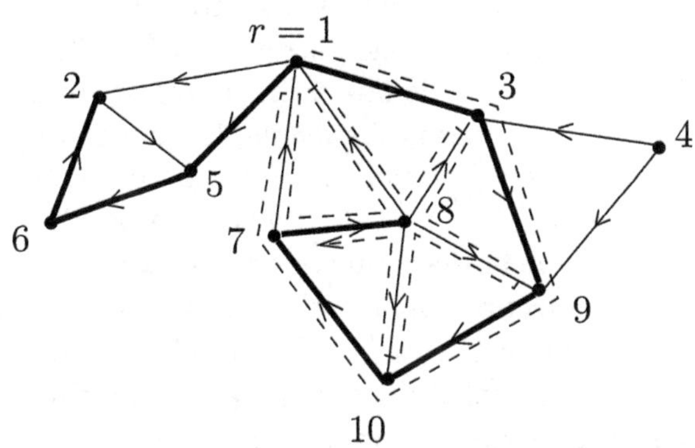

Figure 5.4. *Beginning of a depth-first search of a digraph starting at vertex $r = 1$. The visits are: previsits of 1, 3, 9, 10, 7, revisit of 1, previsit of 8, revisits of 1, 3, 9, 10, postvisit of 8, etc. The arcs in bold are those of previsits (for the complete search). They define in the digraph an arborescence of root $r = 1$, called Trémaux's arborescence. Note that vertex 4 will not be visited by the search*

The sequence of vertices which are in a stack at a given time defines a path in the digraph, called the *current (directed) path* of the search, the last vertex being the *current vertex*, the one where the search is at that time.

5.3.2 Justification

PROPOSITION 5.1. *During a depth-first search of a strict digraph, any vertex accessible by a path from the initial vertex origin r of the search is visited, with a previsit, then a postvisit, and eventually also a revisit.*

Proof. This is easy to justify by reasoning step by step along the vertices from a path of r to the vertex under consideration. □

Note that the vertices which are unreachable by a path from r are not visited at all. We will remedy this later with an *extended* version of the search.

There is an amusing illustration of this algorithm in the study of mazes, the different ways to go through them and, above all, to come out of them! It is in this context that this algorithm has been known for a long time under the name of *Trémaux's algorithm*, named after the author of studies on this subject in the 19th century, long before the theory of the algorithmic graph. Let us imagine that the digraph represents a maze. The vertices represent the crossroads and the arcs the corridors (assumed to be one way). The preceding search defines a systematic exploration strategy: after the entrance, we take a new corridor as long as there is one to take and it leads to a crossroads not yet visited. If no such corridor is available, and if we are back at the entrance to the maze, we stop; if not, we go back to the crossroad where we were before we arrived for the first time where we are now. Of course, to apply this strategy we have to mark one by one every corridor and crossroads followed. There is a famous precedent in Greek mythology with Theseus, who had to find the Minotaur in the labyrinth, to kill him, and then rediscover the entrance to the labyrinth! Ariane was waiting at this entrance with a ball of thread which Theseus unwound during his search. This "Ariane's thread" allowed him to identify the locations previously visited, and, more importantly, to recover the entrance to the labyrinth at the end of his mission. It is interesting to note that the thread corresponds to the state of the stack of the preceding algorithm. More specifically, at each time, the sequence of the crossroads crossed by the thread corresponds, in the digraph, to the sequence of the vertices in the stack (supposing that Theseus was rewinding the thread when he was retracing his steps in a corridor he had already searched).

5.3.3 Complexity

Let us refer to the second iterative version of the algorithm. The main loop is executed, for each vertex v considered, a number of times equal to the outdegree of v, that is $d_G^+(v)$. Each vertex of the digraph is thus considered once. The total number of elementary operations is thus proportional to the sum of the outdegrees of the vertices visited, a sum which is less than or equal to the number m of arcs of the digraph. The complexity of the search itself is thus $O(m)$. To this must be added the complexity required by the initialization of the array `visited`, that is $O(n)$. In total, the complexity is thus $O(\max(n, m))$. The depth-first search algorithm is linear.

We can be even more specific by saying that the search requires a time proportional to the size of what it is visiting (which is not necessarily all of the digraph).

5.3.4 Extended depth-first search

As we have seen, the preceding search only visits the vertices which can be reached by a path from the initial vertex origin r. When we want to visit all the vertices of the digraph, we have to start a new search from a vertex not yet visited, as long as one exists. This search, called an *extended search*, ends when all vertices of the digraph have been visited. This is what is done by the following algorithm, in which it is important to note that the array visited, still supposed initialized to *false* for each vertex, is global with respect to the different search procedures started. This means that it is not reinitialized between successive searches. A vertex is marked visited only once, by one of the searches. In the following expression of the extended depth-first search, the vertices are supposed numbered from 1 to n.

```
procedure dfs_ext(G);
begin
  r:= 1;
  loop
    dfs_ite(G,r);
    -- looking for a non visited vertex
    while r < n and visited(r) loop
      r:= r + 1;
    end loop
    -- r = n or visited(r) false
```

```
    exit when visited(r);
  end loop;
end dfs_ext;
```

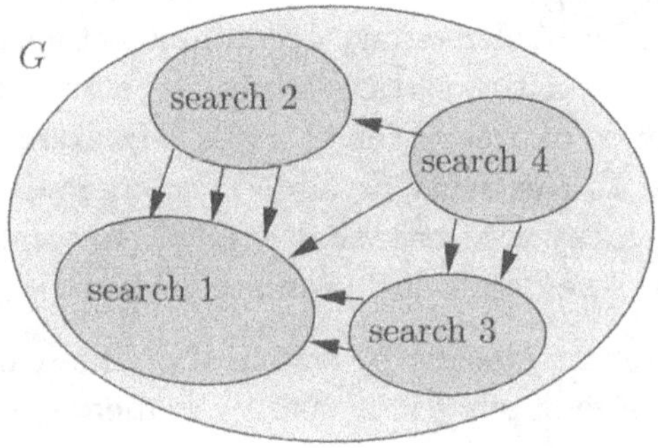

Figure 5.5. *Schema of the successive searches of an extended search*

NOTE. This procedure calls upon the procedure `dfs_ite` but the recursive procedure `dfs_recu` would work just as well.

5.3.5 Justification

PROPOSITION 5.2. *During an extended depth-first search of a strict digraph, each vertex is visited, in a previsit, then a postvisit, eventually also in a revisit.*

Proof. This can be easily justified by observing that any vertex ends up being visited because of the conception of the algorithm itself. □

We can complete the preceding proposition by saying that if the digraph being searched is strict, any arc (u, v) is considered during the search: either during the previsit of v from u, (u, v) is then called a *previsit* arc, or during a revisit of v from u, (v, u) is then called a *revisit* arc. Any arc of the digraph is either a previsit arc or a revisit arc. Note that if (u, v) is a revisit arc, v may have already been visited in a search previous to the one during which u is visited.

5.3.6 Complexity

The various simple searches which constitute the extended one do not overlap, in the sense that a vertex previsited in one of them will not be previsited again in another (let us remember that the array `visited` is global), it can only possibly be revisited. Each search is finite and there is a finite number of searches (at the most equal to the number of vertices, in the case of a digraph without arcs). In addition, as above, each successor of a vertex is considered only once, even when it is a vertex visited in one of the searches with a successor visited in another search. In fact, as we already noted, each search only requires a time proportional to the size of what it is searching, and the total is proportional to the size of the digraph, $\max(n, m)$, thus yielding a linear complexity.

An essential point of the depth-first search, extended or not, is that *when a vertex is postvisited all its successors have been visited, that is previsited from that vertex or revisited (and so previously previsited from another vertex).* This can be clearly seen in particular with the recursive form of the simple (not extended) search. We can also specify that when a vertex is revisited, it has necessarily been previsited (since it is a *re*visit), but it may or may not be postvisited, that is popped or not from the stack of the iterative expression. This point will be useful for recognizing digraphs without circuits.

5.3.7 Application to acyclic numbering

PROPOSITION 5.3. *(1) A strict digraph G is without a circuit if and only if during an extended depth-first search of this digraph, when a vertex is revisited, it has already been postvisited.*

(2) If digraph G is without a circuit, the postorder of the vertices in an extended depth-first search of this digraph is the reverse of that of an acyclic numbering.

Proof. Let us suppose that during an extended depth-first search of digraph G, there is a vertex u which is the successor of current vertex v and which is revisited but not yet postvisited. Vertex u is thus in the stack at that moment, with v, which is at the top of it, and the sequence of the vertices from u to v in the stack is a directed path (directed subpath of the current path of the search). With the arc from v to u, this directed path defines a circuit of G. This proves, by contradiction, the necessary condition of part (1).

Let us now suppose that the preceding circumstances do not happen during a depth-first search of the digraph. So, when a vertex is revisited it is then postvisited. Let us suppose that the vertices are numbered in the reverse postorder, that is during the postvisits, from n to 1. At the time when a vertex v is postvisited, all its successors have been visited, previsited or revisited, according to a general property of the search noted earlier on. Let us show that they are also all postvisited. Those of the successors of v which have been previsited from v are then necessarily postvisited, because they were in the stack above v and thus necessarily popped before v. The other successors of v have been revisited from v. They were then postvisited at the time of their revisit (by hypothesis). Thus, the successors of v were all numbered before v, and since the numbering is done decreasingly from n to 1, they have received a higher number than the one received by v during its postvisit. That is the property which defines an acyclic numbering. It should also be noted that this property is compatible with the extended nature of the search (the numbering is global). The existence of an acyclic numbering under the hypothesis that when a vertex is revisited it is then postvisited, which results itself from the hypothesis without circuits, proves part (2) of the proposition. It also proves the sufficient condition of part (1), because a digraph which admits an acyclic numbering has no circuit, according to proposition 4.1 of Chapter 4. □

5.3.8 Acyclic numbering algorithms

An acyclic numbering algorithm will consist of launching a depth-first search of the given digraph with the following operation while going through the vertices:

- on *postvisit* of v: mark v postvisited and number it decreasing from n to 1.
- on *revisit* of u: if u is not postvisited, stop because there is a circuit in the digraph.

On *previsit* we do nothing.

† As an application, we can again take the digraph in Figure 4.5 in Chapter 4, which is without a circuit, and apply this algorithm to it. (The acyclic numbering obtained may not be identical to the one given on this figure. It can depend on the search followed.)

This acyclic numbering algorithm is, as for the depth-first search, of linear complexity.

5.3.9 Practical implementation

In practice (as we will see later on, in scheduling with the potential task graph), it is not sufficient to detect the presence of a circuit and to stop, because then the digraph has no acyclic numbering. In fact this circumstance corresponds to an error in the datum of the digraph, one arc too many somewhere, creating a circuit. To correct such an error, it is necessary to have a circuit in order to verify its arcs and to detect one that should not be present. The justification of proposition 5.3 makes it possible to exhibit a detected circuit and to do this work, provided that we have access to the search stack. This is direct with the iterative form but is not directly possible with the recursive form, since the stack is managed by the system and therefore hidden. A solution is then to manage an auxiliary stack provided for that effect, pushing and popping respectively on previsits and postvisits as with the stack of the iterative version of the search.

5.4 Exercises

+5.1. Write two programs (in your preferred language) solving the eight queens problem, in the iterative form as well as the recursive, inspiring yourself from the arborescence searches given. Test them (you should find 92 solutions) and compare their time performances. What conclusion can you draw?

5.2. Give an example of a game in which it is the second player who has a winning strategy.

5.3. Consider the following game:[3] two players called A and B choose alternately the digit 1 or 2. A starts. When four digits have been chosen, the game is over and the greatest prime divisor of the number formed by the four chosen digits, written in the order of their choice from left to right, is then what B wins (and what A loses). Explain what may be an optimal strategy for a player and determine this strategy for B, specifying the minimum gain it ensures.

[3]From Boussard-Mahl, *Programmation avancée*, Eyrolles (1983).

N.B. Prime divisor (between parentheses): 1111 (101), 1112 (139), 1121 (59), 1122 (17), 1211 (173), 1212 (101), 1221 (37), 1222 (47), 2111 (2111), 2112 (11), 2121 (101), 2122 (1061), 2211 (67), 2212 (79), 2221 (2221), 2222 (101).

*5.4. *(Digraph kernel and winning strategy)*

Let us go back to the game of Nim 1-3 dealt with as an example, but with a different representation of the game from that of the arborescence associated with it and described in this chapter (Figure 5.3).

a) Find all possible states of the game (there are eight). Build the digraph of which the vertices are these states, with an arc from one vertex to another when it is possible to go from one to the other with one move allowed by the game.

b) Find in this digraph a set N of vertices, called the *kernel*, which has the following properties: for any vertex $x \notin N$, there is an arc joining x to a vertex $y \in N$, arc entering into N, and any arc exiting from a vertex of N has its head outside N. Hint: you will take in N the vertex representing the final state (0-0) of the game.

c) Use the kernel to define a winning strategy (always play by going *into* the kernel).

d) Try to generalize the preceding idea for a winning strategy of any game represented by a digraph which admits a kernel. You will first define the way to play on the given digraph by inspiring yourself from the preceding particular case.

+5.5. a) Show that the previsit arcs of an extended depth-first search of a strict digraph G define a directed forest in G.

b) Try to list the different cases of arcs of revisit which can be found, in particular for the preceding forest (there are three different cases).

Chapter 6

Optimal Paths

Problems of optimal paths, shortest paths or longest paths in graphs or digraphs are diverse and have important applications.

6.1 Distances and shortest paths problems

Here we are considering weighted graphs, most of the time directed. These graphs will be supposed strict when they are directed, simple when they are undirected. Graphs which are not weighted are also dealt with in this chapter, by considering the case when the value of each edge or arc equals 1. The cases of undirected graphs can be brought back to the directed one, simply by replacing each edge by two opposite arcs, each having the same value. In this chapter, directed paths, or walks, in a digraph are simply called *paths* or *walks*.

6.1.1 A few definitions

Let $G = (X, A)$, a digraph which is weighted by a mapping l from A, the set of arcs of G, to real numbers for example. These values represent the *lengths* of the arcs, but in some applications we will see that they can represent many other things: durations, costs, etc.

The *length* of a path, or a walk,[1] of G is defined as the sum of the lengths of its arcs. In the case of a (non-weighted) digraph, with the convention of

[1]Except in section 6.6, in this chapter we always consider paths.

an associated value of 1 for each arc, we again find the length of a path defined as the number of its arcs.

The *distance* from a vertex s to a vertex t in digraph G is naturally defined as the shortest length of the paths from s to t in G. A shortest path from a vertex s to a vertex t is of course a path from s to t of the shortest length possible, that is of length equal to the distance from s to t.

6.1.2 Types of problems

The following problems are formulated for distances. They can be extended for finding corresponding shortest paths as well.

1. *Two vertices problem*: given two vertices s and t, determine their distance.
2. *Single source problem*: given a vertex r, determine the distance from r to each vertex of the digraph.
3. *All ordered pairs problem*: determine the distance from s to t for all ordered pairs (s, t) of vertices.

These problems are not independent. In fact, from problem 1 to problem 3 we encounter problems which are more and more general. Problem 2 is central since it makes it possible to solve problem 3 in a reasonable manner, by repetition from each vertex. It also solves problem 1 in a natural way. Indeed, a direct resolution of problem 1 is not easy. It is simpler to look for the distances one after the other from s, that is implementing a solution of problem 2 with $s = r$, and stopping as soon as vertex t is reached.

In what follows, we will mostly consider problem 2 in specific classic cases.

6.2 Case of non-weighted digraphs: breadth-first search

The length of a path here is simply the number of arcs. Considering problem 2, we want to determine the distances $d(r, s)$. We have $d(r, r) = 0$, the zero length path from r to itself. From there, the calculation principle is the following: if the distances sought after are already determined for all the vertices at a distance $\leq k$, and if a vertex t is of a distance not yet known (therefore $> k$) and successor of a vertex s of which the distance is k, then the distance of t is equal to that of s incremented by 1, that is $k + 1$. This principle is easy to verify. †

This calls for a search strategy of the vertices of the digraph by *successive waves* of vertices. Each wave corresponds to a given distance. The important point of this search is to explore all successors of the vertices of a wave *before* continuing the exploration from a vertex of the following wave. This is the principle of another classic search of graphs, different from the one seen in the preceding chapter, called a *breadth-first search* (abbreviated to bfs), and described below.

The digraph is always given by lists of successors. Array `visited`, indexed on the vertices, is assumed to be initialized to the value *false* for each vertex. This algorithm is copied from the one in the preceding chapter, `dfs_ite`, with the important difference that instead of a *stack* S, we use a *queue* Q. Remember that a queue works under the "First In, First Out" (FIFO) principle (where for the stack the principle is "Last In, First Out", LIFO). The queue primitives used here, `enqueue`, `dequeue`, `front`, `is_empty` are classic. So, `enqueue(Q,t)` puts vertex `t` in queue `Q`, at the rear, `front(Q)` returns the vertex which is at the front of queue `Q` (without removing it from the queue), `dequeue(Q,t)` removes the vertex which is at the front of queue `Q` (without returning it), and `is_empty(Q)` is a function which returns *true* or *false* depending on whether queue `Q` is empty or not. The use of a queue rather than a stack completely changes the search strategy: a vertex which has just been visited is not then immediately used to visit other ones as in `dfs_ite` (Chapter 5), but waits for its turn, stored in the queue. Note that in the expression of the algorithm, the current vertex is not changed when a new visited vertex is added to the queue. This means that there is no need for the assignment `s:= t` after `enqueue(Q,t)` as in the iterative expression of the depth-first search after `push(S,t)` (see Chapter 5).

In addition, it is not necessary, as in the depth-first search, to make a distinction between previsits and postvisits. The nature of the visits to the vertices is specified as commentaries: a first visit, simply called *visit*, and *revisit* in the case of vertices previously visited and met again in the search, as successors of the current vertex.

```
procedure bfs(G,r);
begin
  -- visit of r
  visited(r):= true;
  enqueue(Q,r); s:= r;
  loop
    while suc(s) ≠ null loop
      t:= suc(s).som; suc(s):= suc(s).suiv;
      if not visited(t) then
        -- visit of t
        visited(t):= true;
        enqueue(Q,t);
      else
        -- revisit of t
        null;
      end if;
    end loop;
    deqeue(Q);
    exit when is_empty(Q);
    s:= front(Q);
  end loop;
end bfs;
```

As with the depth-first search, the breadth-first search visits any vertex reachable from r. This is based on the fact that once a vertex has been visited, all its successors are then visited. Thus, along a path from r to a vertex s, each vertex will be visited, in particular s itself.

The complexity of this algorithm is linear. Indeed, the number of elementary operations itself is kept to a constant by the sum of the outdegrees of the vertices, that is the number of arcs m of the digraph. We must also count the initialization of the array `visited` with value *false* for each vertex, initialization which requires a time proportional to the number of vertices n, that is $O(n)$. On the whole, the complexity is thus $O(\max(n, m))$.

6.2.1 Application to calculation of distances

The principle of calculating distances $d(r, s)$ in G is based entirely on this search. Specifically, and by application of the principle stated above, any vertex newly visited is at a distance r equal to that of the current vertex of which it is a successor, incremented by 1.

Thus, the distances may be calculated step by step. The calculation algorithm consists simply of rewriting the preceding algorithm with instead of:

```
-- visit of t
```

the statement:

```
d(r,t):= d(r,s) + 1;
```

and, at the beginning, instead of:

```
-- visit of t
```

the statement:

```
d(r,r):= 0;
```

and nothing when revisiting a vertex, that is in place of:

```
-- revisit of t.
```

6.2.2 Justification and complexity

Let D_k, for integer $k \geq 0$, be the set of the vertices of G at a distance k from r, and let M_k be the set of the vertices marked at a distance k from r by the algorithm. We need to show that $M_k = D_k$ for any k from 0 to the last considered value of k. Note that the sets M_k are pairwise disjoint as well as the sets D_k. Let us reason by induction on integer k. For $k = 0$, the property is clear since both sets D_0 and M_0 are reduced to singleton $\{r\}$. Let us therefore suppose that this property is true for any integer $\leq k$, k being ≥ 0. Let $t \in M_{k+1}$. The vertex t is at a distance $> k$ by the induction hypothesis: otherwise, we would have $t \in M_l$ for $l \leq k$. In addition, according to the way the algorithm works, vertex t has been marked $k+1$ as a successor of a vertex s itself marked k, that is we have $s \in M_k$. Let us consider a shortest path from r to s, of length k augmented by arc (s, t): this is a path from r to t of length $k+1$. This proves that t is at a distance $\leq k+1$. We therefore have $t \in D_{k+1}$, and finally the inclusion $M_{k+1} \subseteq D_{k+1}$, since t is any vertex of M_{k+1}.

Conversely, let $t \in D_{k+1}$. Vertex t being at a distance $k+1$ from r, there is a path from r to t of length $k+1$. Let us consider such a path, and let s be the vertex which just precedes t in it. The distance from r to s is k, since on the one hand the subpath from r to s is of length k, therefore this distance is less than or equal to k, and on the other hand, if it was strictly less than k, the distance of t would be strictly less than $k+1$, which is contrary to the hypothesis. Thus, $s \in D_k$, and since by the induction hypothesis $D_k = M_k$, we have $s \in M_k$, that is, vertex s is marked k. As a successor of s, vertex t will be marked $k+1$, unless it had been previously marked with a value $< k$. However, this is again impossible due to the induction hypothesis, since then we would have $t \in M_l$ with $l \leq k$ and therefore $t \in D_l$, which would contradict $t \in D_{k+1}$. This completes the justification of the algorithm.

As for the complexity, it is of course the complexity of the search, that is $O(\max(n, m))$, also taking into account a possible initialization of the distances. This initialization is useful because it makes it possible to spot the vertices which have not been reached by the search and which may be considered to be at an infinite distance from r. It is also possible to give initially, for each vertex s, the value ∞ to $d(r, s)$, a value which will remain as such for any vertex not reached by the search.

6.2.3 Determining the shortest paths

It is generally interesting to consider with the distances all or some of the shortest corresponding paths. For this, all that is needed is to record for each vertex t, which has just been marked in the algorithm, vertex s to which it succeeds and by which it has been marked, a vertex which we will call its *parent* for the marking of vertices.

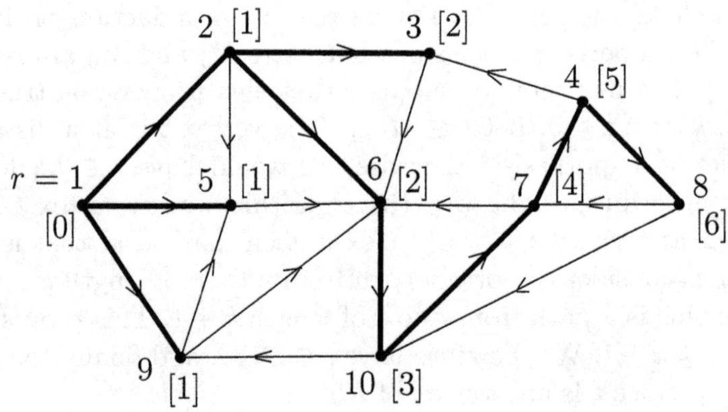

Figure 6.1. *The distances from vertex r are indicated between brackets. The arcs in bold define a shortest path arborescence*

In concrete terms, this can be done with the algorithm previously expressed by associating with the assignment:

```
d(r,t):= d(r,s) + 1;
```

the statement:

```
parent(t):= s;
```

where `parent` is an array indexed on the vertices.

Note that vertex r has no *parent* in that sense. We will have a shortest path from r to any vertex s of the digraph by ascending from s from *parent* to *parent* until r. The shortest path thus obtained will be described by following the sequence of its vertices in the reverse order. The set of the shortest paths thus determined defines in the digraph *a shortest paths arborescence*, of root r. In it, the unique path from the root to any vertex s is a shortest path from r to s in the digraph. Figure 6.1 gives an example.

NOTE. The breadth-first search is not unique and neither is the method of marking the distances. If the distances obtained are, of course, unique, the paths will not be and what is determined here are *some* shortest paths, not *all the* shortest paths in general. Indeed, we must mention that there can be several shortest paths from one vertex to another in a digraph, and this is not rare in applications. For example in the case in Figure 6.1 there are three shortest paths from vertex 1 to vertex 6.

6.3 Digraphs without circuits

We now consider a strict digraph $G = (X, A)$ weighted on its arcs by a mapping $l : A \to \mathbb{R}$, which defines what we call the *lengths* of the arcs. Even though they are interpreted as lengths, the values of the arcs may be negative, which will be useful later on. We will suppose in addition that G has no circuit. Let us note in passing that the absence of circuits makes it possible to avoid some potential difficulties of definition of distances. Indeed, if there is a circuit in a weighted digraph for which the sum of the lengths of the arcs is strictly negative, then the distance from one vertex to another is no longer clearly defined. It is enough to consider, for example, the distance from a vertex of the circuit to itself: this distance is strictly negative, considering the circuit as a path from this vertex to itself, but, furthermore, this distance can be made as small as we want by searching the circuit an arbitrary number of times.

Digraphs without circuits have a characteristic property (given in Chapter 5), which is very useful here, namely the *acyclic numbering* of vertices. Let us recall that acyclic numbering of the vertices is such that if there is an arc from a vertex s to a vertex t, then the number of s is less than that of t. Therefore, when considering the vertices of a digraph in the order of an acyclic numbering, we have the property that when a vertex is considered, all its predecessors have already been previously considered. It is this property which allows the calculation step by step of distances from a given vertex r chosen as the source of distances.

Let us suppose that the vertices of G are indexed in the order of an acyclic numbering, $s_0, s_1, \ldots, s_{n-1}$, with $s_0 = r$. Note that we suppose implicitly that vertex r is a source vertex of the digraph, which we can assume without losing generality since only the vertices accessible by a path from r will be considered for this distance calculation. We have the following formulae:

$$d(s_0, s_0) = 0$$

which results from the fact that the digraph is without circuits, and, for $i = 1, \ldots, n-1$:

$$d(s_0, s_i) = \min_{(s_j, s_i) \in A} \big(d(s_0, s_j) + l(s_j, s_i)\big)$$

where $l(s_j, s_i)$ designates the value (length) of the arc (s_j, s_i).[2] Note that we necessarily have $j < i$ because of the indexing of the vertices following an acyclic numbering. This is what makes this calculation possible, since when we calculate $d(s_0, s_i)$ we know the $d(s_0, s_j)$ (with then $j < i$) which appear in the expression of $d(s_0, s_i)$.

It is easy to transform these formulas into algorithmic form. The algorithm obtained, which we will not formalize further, with a preliminary acyclic numbering algorithm (see Chapter 5), is linear in complexity and efficient practically. Its justification is based on the justification of the preceding formulas. This justification is based in turn on the trivial observation that any shortest path from s_0 to s_i, for $i > 0$, contains as the second last vertex a predecessor of s_i. It is enough then, reasoning by induction on i, to give to this predecessor the role of s_j in the preceding general formula.

[2]Formally, we should write $l((s_j, s_i))$.

6.3.1 Shortest paths

It is possible to obtain some shortest paths corresponding to distances by recording, as with unweighted digraphs, the *parent* of each vertex in determining its distance. It is in fact a vertex s_j for which the *min* of the preceding formula has been reached. A shortest path is then obtained by going back from s_i from parent to parent until vertex s_0.

6.3.2 Longest paths

Shortest paths have numerous applications including relatively unexpected questions *a priori*, such as optimal stock management. However the application chosen in this chapter, particularly important in practice, pertains, in an even more unexpected way, to *longest paths* in weighted digraphs without circuits. A longest path from a vertex s to a vertex t is obviously a path of maximum length linking s to t, called here the *greatest distance* or, rather, *greatest length*[3] from s to t, the maximum of the lengths of the paths from s to t, a quantity denoted here by $D(s,t)$. Note that this concept is correctly defined because there are no circuits in the digraph (especially here, no circuit with positive length).

A simple change of the valuation mapping of the digraph into its opposite, $-l$ instead of l, takes us from the distances to the greatest lengths (up to the positive and negative sign, following the principle that $\max(a,b) = -\min(-a,-b)$), and from the shortest paths to the longest. This change is possible under the express condition that the values of the arcs may be positive, negative or zero. This is true here since no hypothesis has been made above concerning the values taken by l.

6.3.3 Formulas

All that is needed is to replace "min" by "max" in the formulas which express the distances d above, which gives:

$$D(s_0, s_0) = 0$$

[3]The term *length* is to be preferred because *distance* contains the idea of minimum rather than maximum.

and, for $i = 1, \ldots, n-1$:

$$D(s_0, s_i) = \max_{(s_j, s_i) \in A} \big(D(s_0, s_j) + l(s_j, s_i)\big)$$

Determining the longest paths corresponding to these greatest lengths is obtained as in the preceding case by recording the vertices for which the *max* are reached.

The application to scheduling which we are now going to deal with will give some practical examples.

6.4 Application to scheduling

Let a project be broken down into n elementary tasks, numbered 1 to n. For convenience, in the following we will identify a task and its number. A duration is given for each task (in a time unit which may be very different depending on the nature of the project, from the split second to the day), and the *previous* tasks, that is the ones which must have ended before this one can begin. The first question, pompously called the "central scheduling question", is that of the minimum time for completion of the project. The more general problem is that of the planning in time of the different tasks, which we call a *scheduling.* We are going to model the problem with a digraph in which we will apply what we have just seen. The classic method presented here is called the *Critical Path Method.*

6.4.1 Potential task graph

The *potential task graph* is defined as follows. The vertices of this digraph, $s_0, s_1, \ldots, s_n, s_{n+1}$, correspond:

- for $i = 1, \ldots, n$, to the given tasks, s_i corresponding to the task i,
- for $i = 0$ and $i = n+1$, respectively to two fictitious tasks traditionally denoted by α and ω, with duration equal to zero and respectively representing the beginning and the end of the project.

The arcs of the potential task graph are:

- the ordered pairs (s_i, s_j) such that $i, j \in \{1, \ldots, n\}$ and such that the task corresponding to s_i is a previous task of the one corresponding to s_j,

– the ordered pairs (s_0, s_i) for any s_i corresponding to a task without a previous task in the project, and the ordered pairs (s_i, s_{n+1}) for any s_i corresponding to a task which is not a previous task of any task.

Finally this digraph is weighted by associating with each arc (s_i, s_j) the duration of the task corresponding to s_i, which is denoted by d_i. In particular, we have $d_0 = 0$, as the duration of task α.

6.4.2 Earliest starting times

The *earliest starting time* of a task is the earliest time at which this task may begin, the origin of the times being defined by the (instantaneous) completion of fictive task α. In particular, the earliest starting time of ω is the *minimum duration* for completion of the project. The determining of the minimum time for completion is therefore a simple particular case of determining the earliest starting times.

Let us denote by t_i the earliest starting time of task i, for $i = 0, \ldots, n+1$. We have, by definition, $t_0 = 0$. To simplify, we will write from now on $j \prec i$ to indicate that task j is a previous task of task i, a case which corresponds to an arc (s_j, s_i) in the potential task graph. The constraints of the problem are expressed in the following manner, for $i = 1, \ldots, n+1$:

$$t_i \geq t_j + d_j \text{ for all } j \prec i$$

from which we obtain:

$$t_i \geq \max_{j \prec i} \left(t_j + d_j\right)$$

It is essential to note that *the potential task graph is without a circuit.* Indeed, a circuit in this digraph would correspond to a sequence of tasks for which each of them, except for the first one, would have to wait to begin for the preceding one to be over, and the first one would have to wait for the last one to be over. It is obvious that the completion of these tasks is impossible. The potential task graph being without circuits, it allows acyclic numbering. We are supposing from now on that *the tasks are numbered, and the vertices of the potential task graph are indexed, following an acyclic numbering.* As the unique "source" task, α is necessarily task 0 and, likewise, as the unique "sink" task, ω is necessarily task $n+1$. We have the general implication: $j \prec i \Rightarrow j < i$. The expression $\max_{j \prec i}(t_j + d_j)$ of the preceding inequality can then be calculated step by step for i from 0 to $n+1$, thanks to the acyclic numbering. Then, t_i being by definition a minimum, it leads to the equality:

$$t_i = \max_{j \prec i} \left(t_j + d_j\right)$$

This expression is in fact the one which gives the greatest length from s_0 to s_i in the potential task graph, following the general formula given above (see section 6.3.3). In particular, the minimum duration for completing the project is equal to the greatest length from s_0 to s_{n+1}, that is from α to ω, identifying these tasks and their vertices in the potential task graph. Therefore, the minimum duration for completion of the project is equal to $D(\alpha, \omega)$.

6.4.3 Latest starting times

The *latest starting time* of a task is the latest time at which this task must begin for the minimum duration of the project completion to be respected. Let us denote by T_i, for $i = 0, \ldots, n+1$, the latest starting time of task i. The constraint on the minimum duration for the completion of the project is expressed by the equality:

$$T_{n+1} = t_{n+1}$$

† The latest starting times of the other tasks can be calculated by a countdown from T_{n+1}, which leads us to write for $i = n, n-1, \ldots, 1, 0$, in this order:

$$T_i = \min_{j \succ i} \left(T_j - d_i\right)$$

The expression *min* can be calculated step by step in the inverse order of the acyclic numbering, that is for i from n to 0. In addition this expression can, as the one for the earliest starting time, be interpreted as a greatest path length. More specifically, we have the following equality (its justification is proposed as an exercise at the end of this chapter), for $i = n+1, n, \ldots, 1, 0$:

$$T_i = D(\alpha, \omega) - D\left(s_i, \omega\right)$$

6.4.4 Total slacks and critical tasks

The quantity $m_i = T_i - t_i$ is called the *total slack* of the task i. It is the period of time during which we must choose to start a task without pushing back completion of the project, that is by keeping it to its minimum duration. For obvious reasons, a very important practical role is played by *critical tasks*,. those for which $m_i = 0$, that is for which the earliest starting time equals the latest starting time. It results from what we saw above, that these tasks correspond in the potential task graph to the vertices which are on a longest path from α to ω, such a path being called for this reason a *critical path*. The proof of the equivalence of these two concepts (total slack zero and belonging to a critical path) is not completely direct (this is proposed as an exercise at the end of the chapter).

6.4.5 Free slacks

The total slack of a task presents the serious practical drawback that if it is used then some later tasks can become critical. To avoid this, we have to consider a smaller slack. Let us first set out the following expression of a latest starting time calculated not relative to the latest starting time of the following tasks, but relative to the earliest starting time, so for $i = n, n-1, \ldots, 1, 0$:

$$T_i' = \min_{j \succ i} (t_j - d_i)$$

What is called the *free slack* of the task i, is then defined by:

$$m_i' = T_i' - t_i$$

This slack does not have the drawback mentioned for total slack, and is much more useful in practice. In fact, this slack can be used without consequences for the following tasks in the project, that is without delaying them in any way.

NOTE. Generally for any i, we have:

$$0 \leq m_i' \leq m_i$$

and these inequalities can be strict.

Figure 6.2 gives an example of an application of what we have seen. The potential task graph is given directly here for a project with tasks named

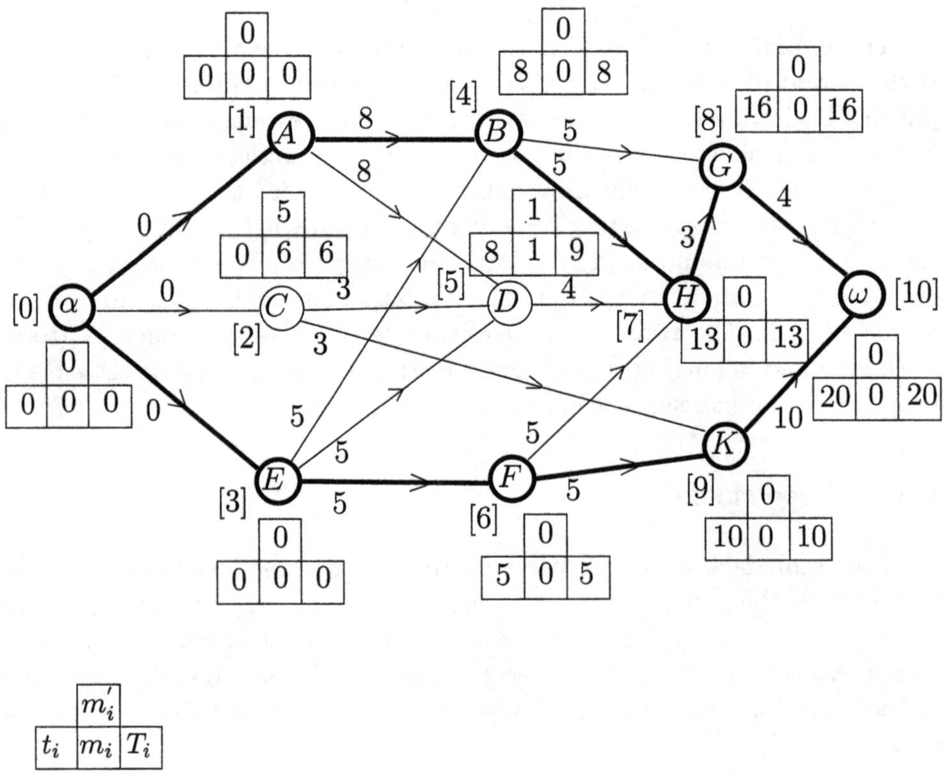

Figure 6.2. *An application of the Critical Path Method*

$A, B, C, D, E, F, G, H, K$. Acyclic numbering of the vertices is given between brackets. This numbering results from the application of a depth-first search of the digraph starting from α. There are two critical paths, in bold in the figure. Task C has a *free* slack strictly contained between 0 and its total slack.

Note that in general, depending on the data of a scheduling problem, the potential task graph is defined by lists of predecessors. In the application of the depth-first search and the determining of the acyclic numbering, we can either convert the list of predecessors into lists of successors or directly proceed in the converse digraph (obtained by reversing the direction of the arcs) starting from vertex ω.

6.4.6 More general constraints

In practice, it is possible to meet constraints such as: task B can only begin after the completion of at least half of task A. Admitting that the degree of completion of a task is proportional to time spent completing it, it is simple to represent such a constraint in the potential task graph by weighting arc (A, B) by half the duration of A. Another fraction of task A, a third for example, would be dealt with in the same way.

Another example: task B must wait a certain time t after the end of task A, the time needed, for example, for paint to dry. All it takes is to weight arc (A, B) by $d_A + t$, where d_A is the duration of A. Thus, in particular, concerning fictive task α, we have the case of a task B which can only start after a given time d, the condition of which is interpreted in the potential task graph by an arc from α to B with value d.

More generally, the potential task graph, and the calculation of the starting times and of the slacks, can take into account any constraints by defining a waiting period between two tasks, which can be different from the duration of the first task as we initially defined.

6.4.7 Practical implementation

Before applying the preceding method, a lot of work has to be done analyzing the project: breaking it down into elementary tasks, determining the duration of each task and the anteriority constraints. The quality of this initial work will strongly influence what will come after. Once this analysis has been done, first schedulings can be calculated, which will lead to the revision of certain data, for example correcting the anteriority constraints which create a circuit in the potential task graph, making the project impossible to complete.

Today, software exists for all types of computers for the calculations and recalculations necessary for this process and the follow up of the completion afterwards. It is possible to edit real dashboards of the project, for example the Gantt chart. In this chart, each task is represented by a line segment placed horizontally where the task in question can be executed following a horizontal temporal axis (in abscissa). The chart is said to be "slanted to the left" when each segment representing a task has its left endpoint placed on the earliest date of this task, and a planning "slanted to the right" when each segment is placed at the latest starting time. This scheduling model makes it possible for the project manager to visualize the progression of the project through time. The critical tasks are indicated on the chart in a specific visible way, to ensure a specific treatment and avoid a delay in any one of them, which would, as we saw, lead to a delay for the entire project.

If we add to the project some constraints with regard to resources necessary for completion of the tasks, and with quantities which are necessarily limited, the problem becomes algorithmically more difficult. The preceding method is not enough because at a given time the tasks which are being executed may require one of the resources in a quantity greater than that available for this resource at that time. For this general problem, we have as data the resource constraints: for each task and each resource, the quantity of this resource which is necessary for completion of one unit of the task. This problem is difficult (**NP**-complete) and its practical importance (resource constraints are present in almost any project) has led to seeking approximate solutions which are as good as possible.

6.5 Positive lengths

Let $G = (X, A)$ be a strict digraph weighted by the mapping $l : (s, t) \in A \to l(s, t) \in \mathbb{R}^+$, that is weighted with *positive or zero* real numbers which here represent *lengths*. Let s_0 be a vertex of G, a vertex from which we will seek the distances and shortest paths. We are still dealing here with the so-called *single source problem* (problem 2 of 6.1.2). Let us put $n = |X|$.

The classic Dijkstra's algorithm, given below, considers the vertices in an order which depends on values on these vertices calculated step by step as work progresses. These values, called *labels*, are defined by the mapping $\lambda : X \to \mathbb{R}^+ \cup \{\infty\}$, and will be the distances sought at the end. The value ∞ represents an infinite distance, which means that a vertex is inaccessible. For computing, it may be represented by a numeric value greater than any value considered. At the beginning, labels are initialized to a value ∞ and are later revised downward step by step. In what follows `t` and $\texttt{t}_0$ are variables of the vertex type and `S` represents an auxiliary set of vertices.

```
procedure Dijkstra1(G,s0);
begin
  -- initialization
  for s ∈ X loop λ(s):= ∞; end loop;
  λ(s0):= 0; S:= {s0}; t0:= s0;
  -- treatment
  while |S| < n loop
    -- revision of labels
    for t ∈ X \ S such that (t0,t) ∈ A loop
      λ(t):= min(λ(t), λ(t0) + l(t0,t));
    end loop;
    -- choice of a minimum label vertex
    t0:= a vertex of X \ S such that λ(t0) is minimum;
    S:= S ∪ {t0};
  end loop;
end Dijkstra1;
```

6.5.1 Justification

This algorithm ends after a finite number of operations, to be specific after $n - 1$ iterations of the main **while** loop (remember that n is the number of vertices of the digraph). Indeed, at each iteration the auxiliary set S increases by one element and ends up equal to X. In addition, each iteration of the main **while** loop clearly only requires a finite number of operations. This number of operations is upper bounded for the **for** loop of labels revision by the outdegree of the arc exiting out of vertex t_0 previously added to S.

Let us then show that at the end of the execution the values of λ are the distances sought after, that is the $d(s_0, t)$ for $t \in X$. Specifically, we are going to show that the equality $\lambda(t) = d(s_0, t)$ for $t \in S$ is initially true and remains such throughout the iterations of the main loop of the algorithm (this property is what is called *the invariant of the loop* **while**). Initially we have $S = \{s_0\}$ and $\lambda(s_0) = 0$ thus, with $d(s_0, s_0) = 0$, trivially the equality to be proved. Let us then show this equality for vertex t_0 introduced in S during a current iteration, supposing the equality true on S before the introduction of t_0. Thus, we will have the equality $\lambda(t) = d(s_0, t)$ for any $t \in S \cup \{t_0\}$, and, step by step, finally for any $t \in X$.

Let μ be any path from s_0 to t_0 in G (see Figure 6.3), and t_μ be the first vertex of μ met, while going from s_0 to t_0, which is not in S. (Note that t_μ exists because s_0 is in μ while t_0 is not.) Let s_μ be the predecessor of t_μ in μ ($s_\mu = s_0$ is possible). Also let ν be the subpath μ from vertex s_0 to vertex s_μ. We denote by $l(\mu)$ the length of path μ, length defined as the sum of the lengths of its arcs, and likewise for $l(\nu)$. We have the sequence of inequalities:

$$l(\mu) \geq l(\nu) + l(s_\mu, t_\mu) \geq \lambda(s_\mu) + l(s_\mu, t_\mu) \geq \lambda(t_\mu) \geq \lambda(t_0)$$

The first inequality is immediate since the path constituted of v augmented by arc (s_μ, t_μ) is a subpath of μ and since the lengths of the arcs are positive or zero. The second inequality results from the fact that the value $\lambda(s_\mu)$ is, by hypothesis on set S, the distance from s_0 to s_μ and therefore less than or equal to the length of path ν. The third inequality results from the determining of $\lambda(t_\mu)$ in the algorithm (value minimized as work progresses). The last inequality results from the choice of vertex t_0, with a minimum label. From $l(\mu) \geq \lambda(t_0)$ it results that $d(s_0, t_0) \geq \lambda(t_0)$. In addition the quantity $\lambda(t_0)$ is the length of a path from s_0 to t_0. Indeed, depending on the determining of λ in the algorithm, there is a vertex $s_{\mu_0} \in S$ such that:

$$\lambda(t_0) = \lambda(s_{\mu_0}) + l(s_{\mu_0}, t_0)$$

Let μ_0 be a shortest path from s_0 to s_0. Then $\lambda(t_0)$ is the length of path μ_0 augmented by arc (s_{μ_0}, t_0). Thus, the quantity $\lambda(t_0)$ is equal to the distance from s_0 to t_0. This completes the justification of the algorithm.

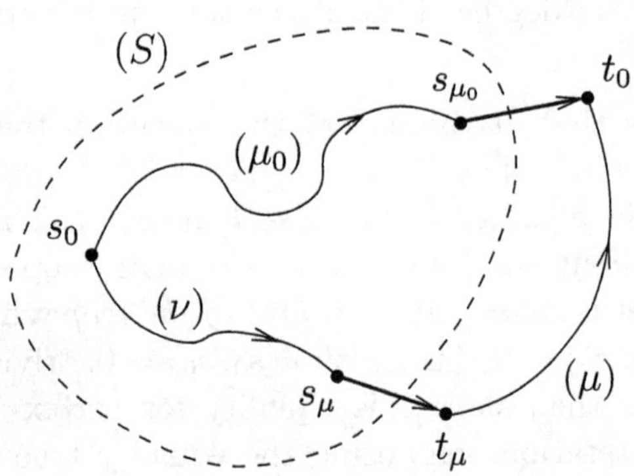

Figure 6.3. *Proof of Dijkstra's algorithm*

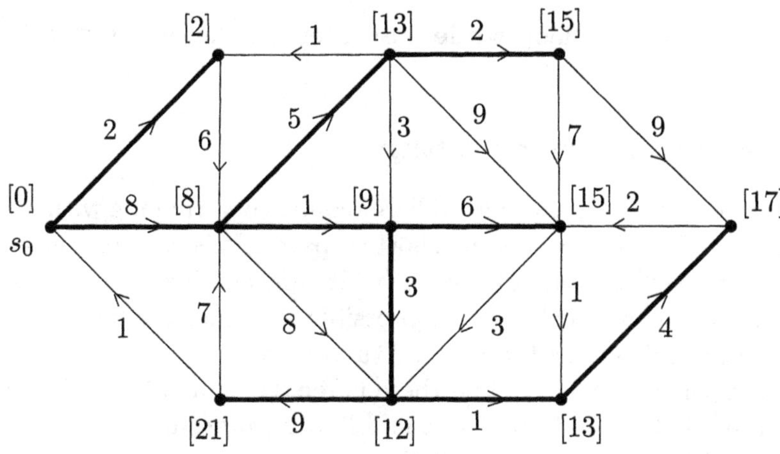

Figure 6.4. *Application of Dijkstra's algorithm (between brackets: the distances obtained, labels in the algorithm; in bold: a shortest paths arborescence)*

An example of an application of Dijkstra's algorithm is given in Figure 6.4.

NOTES. 1) The labels λ help to clarify understanding of the algorithm. With each iteration of the main loop, λ represents for any vertex t the distance from s_0 to t *in the subdigraph of G induced by $S \cup \{t\}$*, that is to say the distance from s_0 to t obtained by considering paths which are only going through vertices of set S. At the end, when $S = X$, we find we are back at the concept of distance defined in G. This makes it possible to understand the role of S during the execution of the algorithm: as the labels of S are definitive, that is they are no longer revised, which means they no longer decrease during the execution of the algorithm, we have $\lambda(t) = d(s_0, t)$ for all $t \in S$.

2) When the next minimum label found is infinite, that is when $\lambda(t_0) = \infty$, it is useless to continue since then vertex t_0 will not improve any other label and in addition all the other vertices of $X \setminus S$ are also in a similar situation (since $\lambda(t_0)$ is a minimum). It is possible therefore to modify the algorithm by adding the instruction:

```
exit when λ(t0) = ∞;
```

at the end of the main loop **while**, just after the choice of vertex t_0 which has a minimum label.

6.5.2 Associated shortest paths

The general problem considered here requires determining with distances all, or some, of the corresponding shortest paths. It is easy to complete the previous algorithm to obtain these paths. It suffices while revising the labels to record the vertex which makes it possible to attain the minimum of the labels. We call this vertex the *parent*. Parents are stored in an array `parent` (indexed on the vertices). To find these parents, we modify the **for** revision loop of the labels as indicated in the following procedure `Dijkstra2`. We have to modify the calculation of the minimum to separate the cases of modification and non-modification of the label. In this second version we also take into account the previous second note concerning the exit of the main loop as soon as a label is infinite.

```
procedure Dijkstra2(G,s0);
begin
  -- initialization
  for s ∈ X loop λ(s):= ∞; end loop;
  λ(s0):= 0; S:= {s0}; t0:= s0;
  -- treatment
  while |S| < n loop
    -- revision of labels
    for t ∈ X \ S such that (t0,t) ∈ A loop
      if λ(t) > λ(t0) + l(t0,t) then
        λ(t):= λ(t0) + l(t0,t);
        parent(t):= t0;
      end if;
    end loop;
    -- choice of a minimum label vertex
    t0:= a vertex of X \ S such that λ(t0) is minimum;
    exit when λ(t0) = ∞;
    S:= S ∪ {t0};
  end loop;
end Dijkstra2;
```

By construction, the vertices which will have at the end of the algorithm a non-zero finite label will have a parent (vertices of S, except s_0). Let $t \neq s_0$ be such that $\lambda(t) < \infty$. Considering the parent of t, then the parent of this new vertex and so on until, necessarily, ending up in the source vertex s_0, we describe, in reverse order, a path from s_0 to t which is a shortest path. Indeed, it suffices to verify that the length of this path is equal to $\lambda(t)$. If † $\lambda(t) = \infty$, there is no path from s_0 to t and so no shortest path (parent of t is not defined).

The shortest paths thus constructed define in G an arborescence with root s_0, an arborescence which is called the *shortest paths arborescence of source* s_0. This arborescence is defined by the primitive *parent* determined in the preceding algorithm. It covers the set of vertices which are at a finite distance from s_0. It is not unique since there may be several shortest paths from one vertex to another. That is why it is said to be an arborescence of shortest paths and not an arborescence of *the* shortest paths.

NOTE. It is often practical to extend to $X \times X$ the mapping l which defines the lengths of the arcs of the digraph under consideration, by putting: $l(s,t) = \infty$ if $(s,t) \notin A$ and $s \neq t$, and $l(s,s) = 0$ for all $s \in X$. In practice, it is convenient to give ourselves the weighted digraph $G = (X, A)$ as a square matrix of order $n = |X|$ with the extended lengths as entries. This is coherent with the initial values given to the labels λ in the algorithm, and with the natural convention on distances that $d(s,t) = \infty$ if there is no path in G from vertex s to vertex t. With these conventions, the **for** loop of revision of labels in the algorithm can be slightly simplified, becoming:

```
for t ∈ X \ S loop
  if λ(t) > λ(t0) + l(t0,t) then
    λ(t):= λ(t0) + l(t0,t);
    parent(t):= t0;
  end if;
end loop;
```

Indeed, in the case $(t_0, t) \notin A$, the instruction applies correctly to $\lambda(t)$, without changing its value since, as $l(t_0, t) = \infty$, we have $\lambda(t_0) + l(t_0, t) = \infty$ and the condition of the **if** loop cannot be fulfilled.

6.5.3 Implementation and complexity

In terms of complexity, Dijkstra's algorithm can first be analyzed in the following way (n still being the number of vertices and m the number of arcs):

- *initialization of labels*: $O(n)$,
- *revisions of labels*: $O(m)$, since there are at the most and in total a number of operations proportional to the outdegree of the arc exiting from each vertex (at the time when this vertex is added to the auxiliary set S),
- *choice of a vertex with a minimum label*: this part of the algorithm requires specifications regarding its implementation. First, and without looking for any refinement regarding the number of operations, it is possible to look for a minimum label by searching all the vertices of $X \setminus S$ and by keeping the smallest label met. The total number of operations is then of the order of the sum of the numbers of the labels considered in each case of a minimum label sought after, that is:

$$(n-1)+(n-2)+\cdots+2+1=O(n^2)$$

In the end, the complexity of Dijkstra's algorithm, as we analyzed it, is therefore:

$$O(n)+O(m)+O(n^2)=O(n^2)$$

(the digraph being strict, the number m of arcs is $O(n^2)$).

This result can be improved, in particular by improving the method for finding a minimum label. Indeed there are classic algorithmic techniques for that (for example, by placing the elements on the nodes of an arborescence called a *priority queue*). Other possible improvements depend on the fact that the digraph is or is not *dense*, that is if the number of its arcs m is of the order n^2 or of a lower order.

6.5.4 Undirected graphs

As we already mentioned in general for problems concerning shortest paths, it is always possible to bring the case of an undirected graph back to that of a directed graph by considering the digraph obtained from the undirected graph after replacing each edge st by two opposite arcs (s,t) and (t,s) having the same value. In practice, we observe that an edge st considered in the revision loop of the labels in the sense of vertex s to vertex t (which corresponds to arc (s,t) in the associated digraph) never has to be considered in the other direction, from t to s. Indeed, with arc (s,t), vertex s plays the role of t_0 in the algorithm. It is therefore in set S, and its label is no longer revizable from t, which means that arc (t,s) will play no role in the revision of labels. Because of this, the previous algorithm can be applied directly to an undirected graph $G = (X, E)$ with only the modification of considering an edge t_0t instead of an arc (t_0, t) in the **for** loop of the revision of labels (version `Dijkstra2`), as follows:

```
for t ∈ X \ S such that t0t ∈ E loop
  if λ(t) > λ(t0) + l(t0,t) then
    λ(t):= λ(t0) + l(t0,t);
    parent(t):= t0;
  end if;
end loop;
```

An example is given in Figure 6.5.

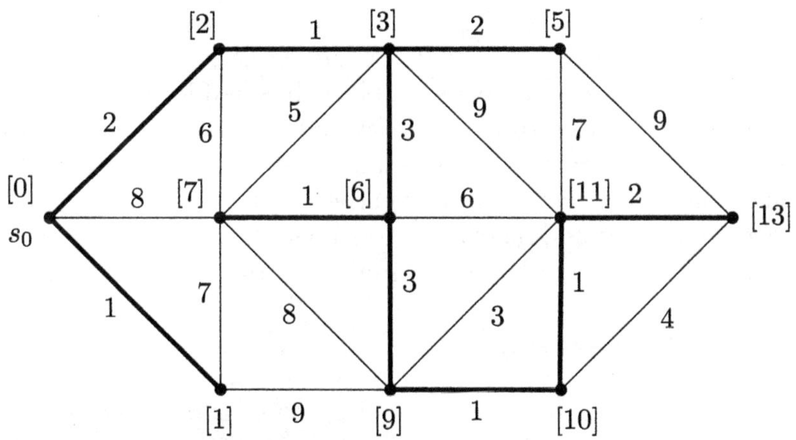

Figure 6.5. *Application of Dijkstra's algorithm to an undirected graph (between brackets: the distances obtained, in bold: a shortest paths tree)*

6.6 Other cases

Dijkstra's algorithm supposes that the lengths of the arcs are positive or zero. This hypothesis is necessary in general, as the example given in Figure 6.6 shows.

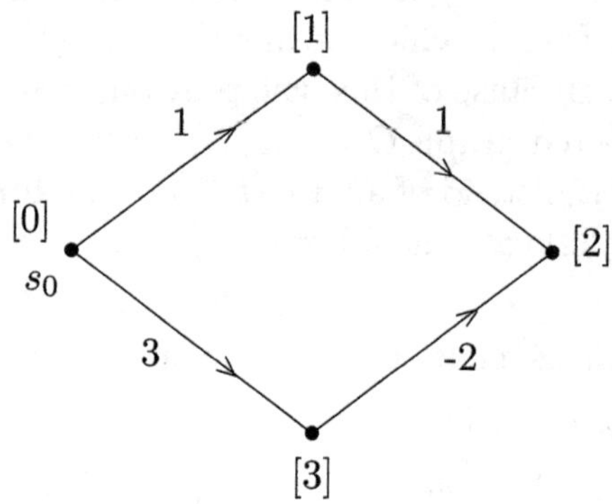

Figure 6.6. *Application of Dijkstra's algorithm to a weighted digraph with a negative arc length. Label 2 found on the vertex on the right is not the distance of this vertex from s_0. There is indeed a path of length 1 going through the bottom vertex*

To escape from this hypothesis is not easy, first of all, as we have observed before, because of the definition itself of distance. Indeed, if in the weighted digraph under consideration there is a circuit with a total negative length, a walk which would meet that circuit could circle it for an indeterminate number of times and would have an arbitrarily small length. We may want to impose a shortest walk to be a path which would prevent it from having to search a negative circuit several times, but it is not easy algorithmically. What must be done in fact, is to preventively detect a negative circuit. For the unique origin problem, there are algorithms which solve this general case, algorithms which are slightly more complicated than Dijkstra's algorithm.

6.6.1 Floyd's algorithm

Still looking at the general case of lengths of arcs which are positive, negative or zero, there is a remarkably simple algorithm for the problem *of all ordered pairs of vertices* (problem 3 of section 6.1.2). Let $G = (X, A)$ be a strict digraph weighted by the mapping $l : (s,t) \in A \rightarrow l(s,t) \in \mathbb{R}$. Let us put $X = \{s_1, s_2, \ldots, s_n\}$ and let us suppose weighted digraph G to be given by the square matrix of order n, $M = (l_{ij})$, defined by:

$$l_{ij} = \begin{cases} l(s_i, s_j) & \text{if } (s_i, s_j) \in A \\ \infty & \text{if } (s_i, s_j) \notin A \text{ and } i \neq j \\ 0 & \text{if } i = j \end{cases}$$

In the following algorithm, `M` represents the adjacency matrix of the digraph and `N` an array variable $n \times n$, which represents the matrix of values calculated step by step through the execution, and which will be at the end, if there are no negative circuits, the distances sought after. Value ∞ is operated classically (for example $\infty + \infty = \infty$). A negative circuit in digraph G is detected by the appearance during the execution of a value `N(i,i)` < 0. If there is no such circuit, at the end of the execution the values obtained `N(i,j)` $< \infty$ are the distances sought after: (`N(i,j)` is the distance from vertex s_i to vertex s_j in G).

```
procedure Floyd(G);
begin
  N:= M;
  for k in 1..n loop
    for i in 1..n loop
      for j in 1..n loop
        if N(i,j) > N(i,k) + N(k,j) then
          N(i,j):= N(i,k) + N(k,j);
        end if;
      end loop;
    end loop;
  end loop;
end Floyd;
```

The justification and the study of this algorithm are proposed in exercise 6.8.

6.7 Exercises

+6.1. Let us go back over the example in section 6.2 (Figure 6.1). Find all the shortest paths from $r = 1$ to the vertex numbered 8 (which is at a distance 6 from r).

+6.2. Let us go back over the example in section 6.4 (Figure 6.2).

a) What happens if we add task G as a previous task to C?

b) Re-evaluate the duration of task E from five to six time units. Show everything that then changes (starting times, slacks).

6.3. A project is broken down into tasks $A, B, \ldots, J$. The following array (at the end of the exercise) gives for each of these tasks its duration and the previous tasks.

a) Draw the potential task graph associated with this project.

b) Calculate the earliest starting times and the latest starting times of the tasks, the total slacks and the free slacks.

c) If task E starts at its latest starting time, are there other tasks which then become critical? If that is the case, say which ones. Repeat the question with tasks A.

d) The constraint that task D cannot begin before starting time 2 is added to the project. Model this constraint in the potential task graph. What does it modify in the preceding results?

Tasks	Durations	Previous tasks
A	13	G
B	9	A, E, I
C	5	B
D	9	—
E	10	D, G, H
F	14	A, E, I
G	5	—
H	6	—
I	14	G, H
J	15	A, D, I

*6.4. Prove the following general expression of the latest starting time T_i, for $i = n+1, n, \ldots, 1, 0$:

$$T_i = D(\alpha, \omega) - D(s_i, \omega)$$

*6.5. Show generally that a task is critical, that is has a zero total slacks, if and only if it belongs to a critical path (that is a longest path from α to ω in the potential task graph).

+6.6. *(Sort by levels the vertices of a digraph without circuits.)*

Let G be a digraph without circuits with a unique source vertex r. Observe that any vertex of G is accessible by a path from r. We will call any set which contains the vertices which are at a same greatest (path) length from r the *level of vertices* in G. In particular, vertex r is at a level corresponding to the greatest length 0. Then, for each greatest possible length, $1, 2, \ldots$ (until the largest value k for which there is in G a vertex with a greatest length k from r), there corresponds a level of vertices in G. The levels, which are so defined and which are not empty, define a partition of the set of the vertices of G, called *sort by levels.* The levels are ordered in the order of the values of their corresponding greatest lengths.

a) Show that two vertices of the same level are never joined by an arc.

b) Show that each arc of G goes from a level to a higher level.

c) Show that by numbering the vertices of G in the increasing order of levels, and randomly within the same level, we obtain an acyclic numbering.

d) Find out the sorting by levels of the vertices of the following digraph:

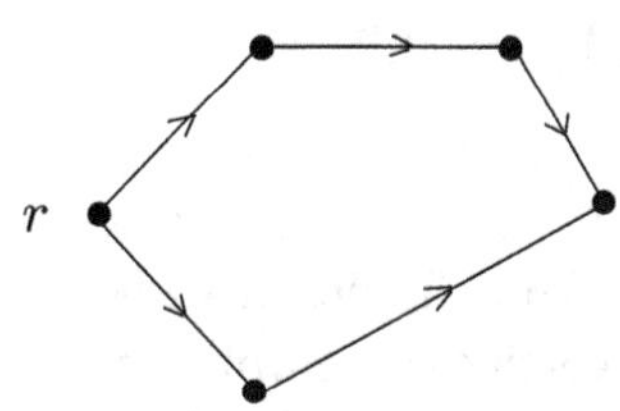

e) Find in this digraph an acyclic numbering which cannot be obtained from a sorting by level obtained as indicated in question c.

6.7. Apply Dijkstra's algorithm to the undirected graph in Figure 6.7 below, taking vertex s_0 as the origin of the distances. Deduce a shortest path from vertex s_0 to vertex s_2.

*6.8. *(Study of Floyd's algorithm)*

Let us consider Floyd's algorithm as it is described in section 6.6.1. Let us put, for $k = 0, 1, \ldots, n$, $d_k(s_i, s_j)$ the shortest length of the

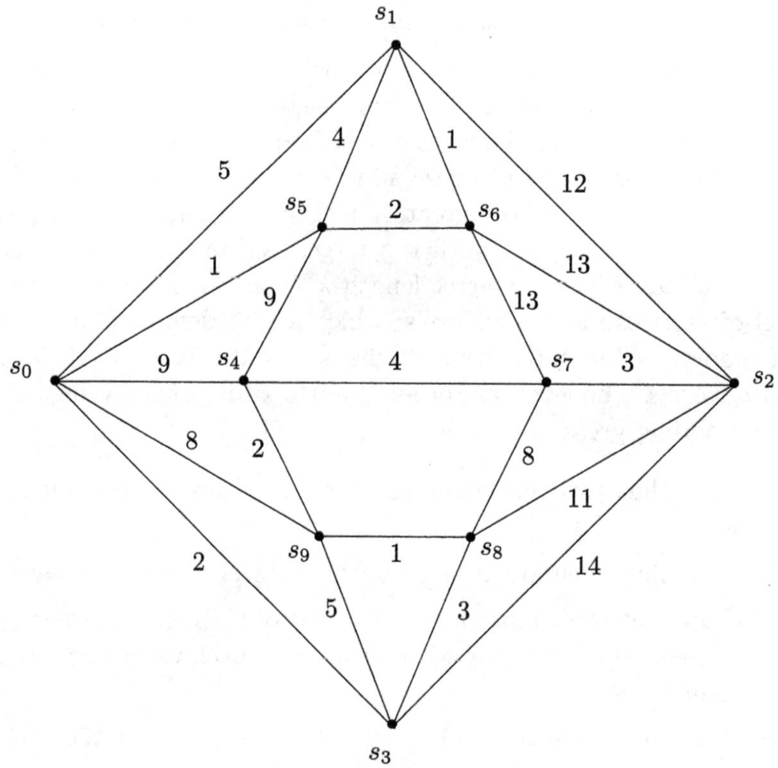

Figure 6.7. *Exercise 6.7*

walks of G going from s_i to s_j and whose internal vertices belong to the set $\{s_1, \ldots, s_k\}$. We call such a walk a *k-walk*. For $k = 0$, the set of possible internal vertices is empty and the 0-walks correspond to the arcs. Let us denote by N_k, for $k = 1, \ldots, n$, the array `N` of the algorithm at the end of the execution of the kth iteration of the loop **for** k `in` $1 \ldots n$ (first **for** loop). At the beginning, N_0 is the adjacency matrix M of the digraph and N_n is the final `N` of the algorithm.

a) Show by induction on k that if $N_k(i, j) < \infty$ then $N_k(i, j)$ is the length of a k-walk from vertex s_i to vertex s_j, this result being true even in the case $i = j$.

b) Deduce from the question above that if there is a $N_k(i, i) < 0$ then there is a negative circuit in G (circuit with a length < 0).

c) Show by induction on k that whatever i, j, k are, $N_k(i,j)$ is less than or equal to the length of any k-path from s_i to s_j (path which is a circuit in the case $i = j$).

d) Deduce from the previous question that if there is a negative circuit in G, then there is a $N_k(i,i) < 0$.

e) It follows from the preceding questions that digraph G will have a negative circuit if and only if a `N(i,i)` < 0 appears during the execution of the algorithm. We suppose that digraph G has no negative circuits. Show that if $N_k(i,j) < \infty$ then $N_k(i,j) = d_k(s_i, s_j)$. Deduce from this, that at the end of the algorithm any value `N(i,j)` $< \infty$ is the distance from s_i to s_j in G.

f) Show how we can recover shortest paths corresponding to the distances found.

g) Study the complexity of Floyd's algorithm. Try to compare, from a complexity point of view, Floyd's algorithm and Dijkstra's algorithm applied to each of the vertices of the digraph taken as source (which will also give the distances for all ordered pairs of the vertices).

*6.9. *(To reflect on)*

Let $G = (X, A)$ be a strict digraph weighted by integers > 0, that is with a mapping $l : A \rightarrow \mathbb{N}^*$. We associate with G the digraph $H = (Y, B)$ obtained by replacing in G each arc $a = (s, t)$ by a path of length $l(a)$ going from s to t. All these paths are in H pairwise disjoint for their internal vertices. Let us call *main vertices* of H those vertices coming from G (the ones which are not main vertices of H are the internal vertices of the preceding paths). We identify the main vertices of H and the corresponding vertices of G. Digraph H is not weighted and the application to this digraph of the distances calculation algorithm from a main given vertex s_0 (section 6.2) gives in particular the distances from s_0 to each of the main vertices of H. These distances correspond, by construction of H, to the distances from s_0 in G. Try to rediscover Dijkstra's algorithm applied to G from the preceding algorithm applied to H.(Observe the order in which the main vertices in H are dealt with. You will find once again the idea of Dijkstra's algorithm of dealing with the vertices *in the order following the minimum labels*).

Chapter 7

Matchings

The concept of matching is one of the major concepts of graph theory. It is of theoretical interest because of its development and its links with other concepts, especially the concept of flow studied in the following chapter. It is also interesting for its applications, in particular for the classic optimal assignment problem dealt with in this chapter.

7.1 Matchings and alternating paths

7.1.1 A few definitions

A *matching* of a graph G is a set M of edges such that no two edges share a same endvertex. A vertex of the graph is said to be *saturated* by a matching M, or M-saturated, if it is an endvertex of an edge of M. Otherwise it is said to be *unsaturated* by M, or M-*unsaturated*. If every vertex of G is M-saturated, then matching M is said to be *perfect*. A matching is said to be *maximal* if it is impossible to add an edge to it. A matching M is said to be *maximum* if it has the greatest possible number of edges.[1]

NOTES. 1) If a graph has an odd number of vertices, it, of course, cannot have a perfect matching. (This is the case for the graph in Figure 1(b)).

[1]Note the difference between "maximal" and "maximum". Generally speaking, maximal is used when for a certain case nothing can be added to make it bigger, while maximum is used when there are no other cases bigger than the maximum. There is the same distinction between "minimal" and "minimum".

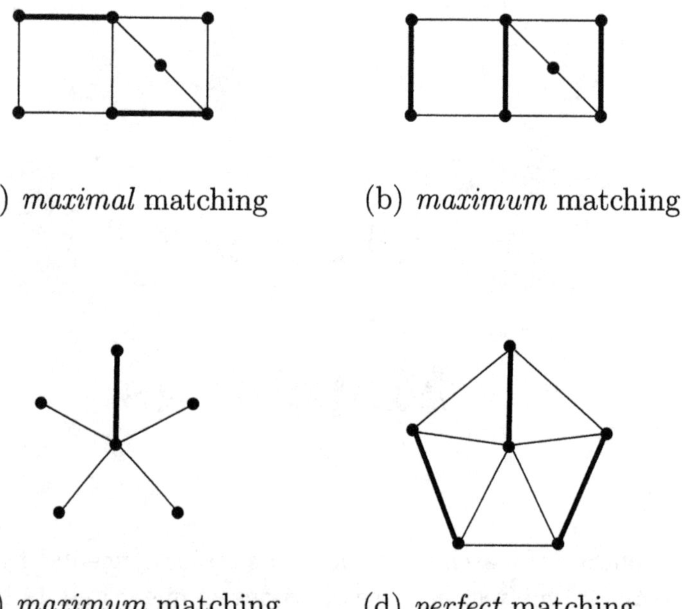

(a) *maximal* matching (b) *maximum* matching

(c) *maximum* matching (d) *perfect* matching

Figure 7.1. *Matchings examples (edges in bold lines). Example (a) shows the case of a matching which is* maximal *but not* maximum

2) If a graph is bipartite, of the form $G = (X, Y, E)$, it can have a perfect matching only if $|X| = |Y|$. This condition is necessary but not sufficient.
† (Find a counter-example with a connected bipartite graph.)

Application of the second note enables us to solve very simply the problem of using dominoes to tile a "truncated chessboard". A chessboard (8×8 squares) has had its right bottom corner square and its upper left corner square removed. The point is to show that it is impossible to cover it exactly with dominoes: a domino covers two squares with a common side, horizontally or vertically. Covering exactly means without leaving any square uncovered and without overlapping dominoes. Such a covering is called a *domino tiling.*[2] The solution is easy if we look at the colors of the squares: each domino necessarily covers a white square and a black one. In the truncated chessboard there are two more black squares than white ones (deciding by convention that the squares removed, necessarily of the same color, are white), which is incompatible with the preceding property. We apply the previous note to the graph associated with the chessboard in

[2]A problem on the study of domino tiling is proposed in Chapter 12.

the following manner: the vertices are the squares of the chessboard, the edges are pairs of squares sharing a common side. It is easy to see that this graph is bipartite (vertices associated with white squares on the one hand, vertices associated with black squares on the other). A matching of the graph corresponds to a tiling of the chessboard. Since the two classes of the bipartition do not have the same number of vertices, there is no perfect matching in the graph and therefore no tiling of the chessboard.

7.1.2 Concept of alternating paths and Berge's theorem

Given a graph $G = (X, E)$ and a matching M of G, an *alternating path* relative to M, or an M*-alternating* path, is a path for which the edges are alternately in M and in $E \setminus M$. An M-alternating path is said to be an *augmenting path* for M, or an M*-augmenting path*, if its endvertices are M-unsaturated.

The following theorem is the foundation of the whole theory.

THEOREM 7.1 (Berge). *A matching M of a graph G is maximum if and only if there is no M-augmenting path.*

Proof. Let us first consider the case of the existence of an M-augmenting path, C. Let x and y be the ends of C, M-unsaturated vertices by hypothesis. Let us define a new matching M' of G as follows: any edge of path C which is not in M is taken in M', as well as any edge of M which is not in C. We say that M' is obtained *by exchanging the edges of* M along path C. It is easy to verify that M' is effectively a matching of G. This matching M' has one edge more than M (compare the number of edges of M and M' in path C). Thus, if an M-augmenting path exists, it can define a matching M' such that $|M'| > |M|$. Matching M is not maximum, which proves the necessary condition of the theorem.

Conversely, let us suppose that matching M is not maximum and let M' be a matching such that $|M'| > |M|$. Let us consider the spanning subgraph of G induced by the edges of G which are in M or in M' but not in both. Each vertex of H is of degree ≤ 2 (in H). It is an easy exercise to show that †
the connected components of such a graph are one of the following: (1) an isolated vertex; (2) a cycle, necessarily even (an *alternating* cycle, its edges are alternately in M and in M'); (3) an alternating path (relative to M and M'). In cases (1) and (2) the number of edges of each of the two matchings

in the component is the same. Only in case (3) can there be a different number of edges in M and in M'. Since there are more edges in total in M' than in M, this must be also be found in one of the components of H, and therefore it must be one of the components which is in case (3). We thus have shown the existence of a path in which the edges are alternately in M and in M', with one edge more in M' than in M. It is easy to see that the ends of such a path are M'-saturated, therefore M-unsaturated. Thus this path is M-augmenting. The existence of this path shows, by the absurd, the sufficient condition of this theorem. □

Going back over the example in Figure 7.1(a), it is easy to find an augmenting path which makes it possible to define, by exchange of edges, a matching having one edge more (for example the matching in case (b)). On the other hand in case (c), for example, there is no augmenting path; this matching is truly maximum.

7.2 Matchings in bipartite graphs

The theory of matchings is general but here we will restrict it to the case of bipartite graphs, which is already interesting and which is enough for the important assignment problem which we will study later on.

Given a bipartite graph $G = (X, Y, E)$ and a matching M of G, we will use the following definitions.

If $xy \in M$, the vertices x and y are said to be *matched under* M, or M-matched, or vertex x is said to be *matched* or *M-matched* to vertex y.

The sets $U \subseteq X$ and $V \subseteq Y$ are said to be *matched under* M, or M-matched, if any vertex of U is M-matched to a vertex of V and vice versa.

The set $U \subseteq X$, or $V \subseteq Y$, is said to be *M-saturated* if all its vertices are M-saturated. It is also said that M *saturates* U or is a *U-saturating* matching.

In what follows, given $S \subseteq X$, $N(S)$ designates the set of the neighbors of the vertices of S in G. We have $N(S) \subseteq Y$.

We are now going to establish a necessary and sufficient existence condition for a matching in a bipartite graph, a result which is fundamental for what is to come. The following lemma will make the proof very simple.

LEMMA 7.1. *Let $G = (X, Y, E)$ be a bipartite graph and let M be a maximum matching of G. Let us put: S_0 the set of the vertices of X which are M-unsaturated, Z the set of the vertices of G which are joined to at least one vertex of S_0 by an M-alternating path, $S = Z \cap X$ and $T = Z \cap Y$ (note that we have $S_0 \subseteq S$). Then: 1) $N(S) = T$, 2) $S \setminus S_0$ and T are M-matched.*

This very technical lemma will be demonstrated by an exercise (in †
considering an example, it is easy to see the way to follow).

THEOREM 7.2 (Hall). *A bipartite graph $G=(X, Y, E)$ allows an X-saturating matching if and only if for any $S \subseteq X$ we have $|N(S)| \geq |S|$.*

Proof. Let us show the sufficient condition. Let M be a maximum matching of G. Using the notations and the results of lemma 7.1, we have, if $S_0 \neq \emptyset$:

$$|N(S)| = |T| = |S \setminus S_0| = |S| - |S_0| < |S|$$

which contradicts the hypothesis $|N(S)| \geq |S|$. Thus $S_0 = \emptyset$ and therefore matching M saturates X. Conversely, let $S \subseteq X$ be such that $|N(S)| < |S|$. Then an X-saturating matching should match the vertices of S to a set of vertices of $N(S)$ of cardinality at least equal to the cardinality of S, but that is not possible since $|N(S)| < |S|$. Therefore such a matching cannot exist. □

The following corollary is a classic result in algebra.

COROLLARY 7.1 (Marriage lemma). *If $G = (X, Y, E)$ is a k-regular bipartite graph, with $k > 0$, then G has a perfect matching.*

Proof. We have $|E| = k|X| = k|Y|$, and thus $|X| = |Y|$. An X-saturating matching is therefore perfect. Let S be any subset of X and let us put F_1 the set of the edges incident to a vertex of S, F_2 the set of the edges incident to a vertex of $N(S)$. The number of edges of F_1 is $k|S|$ and the number of edges of F_2 is $k|N(S)|$. As $F_1 \subseteq F_2$, since any edge incident to a vertex of S is also incident to a vertex of $N(S)$, we have $k|S| \leq k|N(S)|$. This gives $|N(S)| \geq |S|$, which proves the existence of an X-saturating matching by application of theorem 7.2. □

This corollary takes it name "Marriage lemma" from the fact that it solves the following problem: an equal number of boys and girls are to be wed and it is supposed that each boy and each girl agrees to marry someone of the opposite sex among k of them selected by everyone *a priori*. The relation is symmetric (if x selects y, y selects z) and this number k is the same for every boy and every girl. We need to show that it is possible for everyone to marry, each person with one of the persons belonging to his or her selection. The answer, far from obvious, is positive according to corollary 7.1. The problem is modeled using a bipartite graph $G = (X, Y, E)$, where X is the set of boys, Y the set of girls, and E the set of boy–girl pairs which have selected each other for a possible marriage. A perfect matching answers this question.

7.2.1 Matchings and transversals

Matchings are closely linked to the concept of transversal, which is dual in some way. First let us give some definitions concerning general graphs, assumed not to be bipartite.

A *transversal set*, or simply a *transversal*, of a graph G is a set of vertices of this graph such that any edge of G has at least one endvertex in this set.

A *minimum transversal* is a transversal which has the lowest possible number of vertices.

(a) transversal (a) *minimum* transversal

Figure 7.2. *Examples of transversals (vertices in bold)*

The *transversal number* of a graph G is defined as the lowest number of vertices of a transversal of G, that is the cardinality of a minimum transversal. This number is denoted by $\tau(G)$.

The *matching number* of a graph G is defined as the largest number of edges in a matching of G, that is the cardinality of a maximum matching. This number is denoted by $\nu(G)$.

It is easy to see that *in a graph the cardinality of any matching is less than or equal to the cardinality of any transversal.* Indeed, any edge has at least one endvertex in the transversal, and two edges of the matching have no common endvertex. Thus, in general we have:

$$\nu(G) \leq \tau(G)$$

However, we do not always have the equality, as the example of the graph in Figure 7.2 shows, a graph for which $\nu(G) = 3$ and $\tau(G) = 4$. It is remarkable that for bipartite graphs, on the other hand, we always have this equality as we will see. The following result is at the basis of the demonstration of this equality and it is also easy to prove.

LEMMA 7.2. *Let G be any graph, M a matching of G, and L a transversal of G. If $|M| = |L|$ then M is a maximum matching and L a minimum transversal of G.*

Proof. Any matching having a number of elements less than or equal to that of any transversal, if the cardinality of a matching is equal to that of a transversal, the number of elements of the matching is necessarily a maximum and the number of elements of the transversal a minimum. □

LEMMA 7.3. *Let $G = (X, Y, E)$ be a bipartite graph, M a matching of G, S_0 the set of the vertices of X which are M-unsaturated. Let $S \subseteq X$ and $T \subseteq Y$ be sets such that $S_0 \subseteq S$, $N(S) = T$ and $S \setminus S_0$ and T are M-matched. Then $(X \setminus S) \cup T$ is a transversal of G which has the same cardinality as M.*

Proof. The proof of this lemma is easy. First it is simple to verify that $L = (X \setminus S) \cup T$ is a transversal of G, since $N(S) = T$. Then we have, because of the hypotheses on sets S and T:

$$\begin{aligned} |L| &= |X| - |S| + |T| = |X| - |S| + \big(|S| - |S_0|\big) \\ &= |X| - |S_0| = |X \setminus S_0| = |M| \end{aligned}$$
□

The second following theorem is the other major basic result of matching theory in bipartite graphs. It is in addition equivalent to theorem 7.2.

THEOREM 7.3 (König). *If graph G is bipartite, then $\nu(G) = \tau(G)$.*

Proof. Consider a maximum matching of G and sets S and T as defined in lemma 7.1. Note that the hypotheses of lemma 7.3 on sets S and T are the conclusions of lemma 7.1. The linking of lemmas 7.1 and 7.3, with lemma 7.2, justifies the theorem. □

7.3 Assignment problem

In a company, employees must be assigned to positions. The positions he or she may take are given for each employee. The question is to find out whether it is possible to assign all employees, with one employee to each position and vice versa, and, if so, to give such an assignment.

The problem can be modeled using a bipartite graph $G = (X, Y, E)$, where X is the set of employees, Y the set of positions, E the set of employee-position pairs such that the employee can occupy the position considered. Assignment of employees, possibly not all of them, corresponds in graph G to a matching, and assignment of all employees corresponds to a matching of G which saturates X. Theorem 7.2 (Hall) provides an answer to this question. The solution exists if and only if for any set S of employees the number of positions which can be occupied by at least one employee of S is greater than or equal to the number of employees of S.

We are going to look here at a practical algorithmic method for answering this question and give a solution when there is one.

7.3.1 The Hungarian method

Given a matching M which does not saturate X (and which may be initially empty), we look for an augmenting path. This search is made from an unsaturated vertex r of X by a systematic exploration of the alternating paths which start from r, until another unsaturated vertex is found, thus bringing to light an M-augmenting path which makes it possible to augment the matching. We start again with the matching augmented by the preceding path, as long as there is an unsaturated vertex in set X. Let us specify all this in the following algorithm, known under the name of the *Hungarian method.* Graph G is assumed to be given by what we can call a *neighborhood function* N such that for each $S \subseteq X$, $N(S)$ is the set of the neighbors of the vertices of S (it is more convenient here to consider the neighbors of a vertex in a set rather than in a list).

In what follows, I represents the set of the vertices of X which are unsaturated by the current matching M, and the Boolean variable found is used to signal the discovery of an augmenting path.

```
procedure matching_saturating_X(G);
begin
  M:= ∅; I:= X;
  loop
    exit when I = ∅;
    r:= a vertex of I;
    look_for_augmenting_path(M,r);
    if found then
      augment_matching(M,r);
    else
      exit;
    end if;
    I:= I \ {r};
  end loop;
  -- if I = ∅, M is a matching which saturates X
  -- if I ≠ ∅, there is no matching which saturates X
end matching_saturating_X;
```

Some of the preceding instructions are *really* pseudo-instructions which have to be made explicit. Sets S and T (variables S and T) considered in what follows are auxiliary sets of vertices, respectively of X and of Y.

```
look_for_augmenting_path(M,r):
S:= {r}; T:= ∅;
s:= r;
loop
  if N(S) = T then
    found:= false;
    exit;
  end if;
  t:= a vertex of N(S) \ T;
  parent(t):= a vertex of S ∩ N({t});
  if t M-insatured then
    found:= true;
    exit;
```

```
    end if;
    s:= the vertex M-matched to t;
    parent(s):= t;
    S:= S ∪ {s}; T:= T ∪ {t};
  end loop;

  augment_matching(M,r):

  s:= parent(t); M:= M ∪ {st};
  while s ≠ r loop
    t:= parent(s); M:= M \ {st};
    s:= parent(t); M:= M ∪ {st};
  end loop;
```

Considering the `look_for_augmenting_path` the search for a vertex t in $N(S) \setminus T$ and then for a vertex in $S \cap N(\{t\})$, put as a parent of t, note that this can be done in practice in the same movement: vertex t is found as a neighbor of a vertex s of S and this vertex s can then be put as a parent of t in the array variable `parent` of `look_for_augmenting_path`.

Note in addition that `augment_matching` only applies when `look_for_augmenting_path` has been called previously and has given `found = true`. Vertex t, initially considered in `augment_matching`, is the one considered last in `look_for_augmenting_path`. The array `parent` has been filled for the concerned vertices in `look_for_augmenting_path` and it makes it possible, immediately after in `augment_matching`, to follow the found augmenting path to increase the current matching.

7.3.2 Justification

The algorithm ends in both cases of the **exit** instruction of the procedure: $I = \emptyset$ and $N(S) = T$. This second case corresponds to `found = false` in `look_for_augmenting_path`. One of these two cases ends up happening after a finite number of operations. Indeed, on the one hand in `look_for_augmenting_path` sets S and T increase at each iteration, S in X and T in Y. If there is not at any point $N(S) = T$, we end up finding an unsaturated vertex t, which makes it possible to augment the matching. At the other end, each iteration of the loop **loop** of the procedure decreases set I of the unsaturated vertices. This set will therefore end up being empty for lack, meanwhile, of an exit in the other case, $N(S) = T$.

Let us verify the result obtained. The case of the instruction **exit** with $I = \emptyset$ is clear: the current matching M saturates X. Let us consider the case $N(S) = T$. At each step of the construction of sets S and T, in `look_for_augmenting_path`, we have $|T| = |S| - 1$: it is true initially, with $S = \{r\}$ and $T = \emptyset$, and this remains true with each iteration of the interior loop, since sets S and T are always increased together, each with one element. When $N(S) = T$, we therefore have $|N(S)| = |T| = |S| - 1 < |S|$. Thus the set S found contradicts the necessary condition of theorem 7.2 (Hall), which shows that there is no X-saturating matching.

NOTE. Set S given by the algorithm in the case of non-existence of an X-saturating matching is important: it makes it possible to "certify" the answer. It is enough in fact to see that $|N(S)| < |S|$ to be sure of the non-existence of such a matching. We find the concept of a "certificate" in complexity theory (see Appendix B). It is important also from a practical point of view, because in the negative case it matters to be able to verify the answer of the algorithm and possibly to see what brought out this negative answer. Note that the positive answer is certified by the datum of an X-saturating matching. Overall, the existence property of X-saturating matching is said to be "well characterized".

7.3.3 Concept of alternating trees

Let us consider the subgraph of G induced by sets S and T and the edges which, in `look_for_augmenting_path`, define the array *parent*. This subgraph is an arborescence of root r which has the property that the path from r to any vertex of the arborescence is M-alternating. It is called an *alternating tree.* In the case $N(S) = T$, it is called a *Hungarian tree.* As we saw in the preceding justification, we then have $|T| = |S| - 1$. Another property of the Hungarian case is that $S \setminus \{r\}$ and T are M-matched. The existence of a Hungarian tree having an M-unsaturated vertex r means that there is no M-augmenting path having this vertex as its end.

7.3.4 Complexity

Let us put $n = |X|$. The loop **loop** of the procedure `matching_saturating_X` is searched a maximum of n times. Indeed, each iteration of this loop either saturates a vertex of X, and therefore decreases the set of the unsaturated vertices I by a unit, or is the last iteration and is in the exit case with the variable `found` equal to false.

For a given unsaturated vertex r, the interior loop of `look_for_augmenting_path` is searched at the most $n - 1$ times, the maximum of elements which it is possible to add in S. Likewise, the loop in `augment_matching` does not require more than n iterations. Finally the initialization of M and of I does not require more than a time $O(n)$. In the end, this algorithm is polynomial, and has a complexity which may first be evaluated in $O(n^4)$, and then in $O(n^3)$. The polynomial nature of this algorithm is an interesting result in itself. Non-trivial improvements make it possible to obtain much better, in terms of complexity, $O(m\sqrt{n})$ in particular.

NOTE. In practice it is possible to reduce the number of main iterations by starting from a non-empty matching. Indeed, this algorithm works from any matching by initially taking for I the set of the vertices of X which are unsaturated for this matching. By hand, this makes it possible to avoid some first trivial iterations such as an alternating path reduced to one edge and its endvertices.

7.3.5 Maximum matching algorithm

It is easy to extend the previous algorithm in order to get a maximum matching of the graph. It is enough not to stop when condition $N(S) = T$ is reached. All the vertices of X are considered for the search of an augmenting path. Set I is now the set of the vertices of X remaining to be considered.

```
procedure maximum_matching(G);
begin
  M:= ∅; I:= X;
  loop
    exit when I = ∅;
    r:= a vertex of I;
    look_for_augmenting_path(M,r);
    if found then
      augment_matching(M,r);
    end if;
    I:= I \ {r};
  end loop;
  -- matching M is maximum
end maximum_matching;
```

The two pseudo-instructions `look_for_augmenting_path` and `augment_matching` are not modified with regard to the previous algorithm.

7.3.6 Justification

When we have a Hungarian tree with the current unsaturated vertex r (where $N(S) = T$) as its root, we restart the main loop from another vertex of I, if there is still one. Thus, at the end, we have for each vertex r of I: either r is saturated by the matching M obtained, or r is unsaturated and then is the root of a Hungarian tree. Let $r_1, r_2, \ldots, r_q$ be the vertices of X unsaturated at the end of the execution, $S_1, S_2, \ldots, S_q$ and $T_1, T_2, \ldots, T_q$ sets S and T successively built in the algorithm and which define the Hungarian trees having for respective roots $r_1, r_2, \ldots, r_q$. Let us put $\mathcal{S}_0 = \{r_1, r_2, \ldots, r_q\}$, $\mathcal{S} = \bigcup_{i=1}^{q} S_i$ and $\mathcal{T} = \bigcup_{i=1}^{q} T_i$. Since $N(S_i) = T_i$ for $i = 1, 2, \ldots, q$, we have $N(\mathcal{S}) = \mathcal{T}$. Since, on the other hand, $S_i \setminus \{r_i\}$ and T_i are M-matched for each i, $\mathcal{S} \setminus \mathcal{S}_0$ and $\mathcal{T}$ are M-matched. Lemma 7.3, applied to the sets $\mathcal{S}_0$, $\mathcal{S}$, $\mathcal{T}$ (in place of S_0, S, T), and lemma 7.2 make it possible to conclude that M is a maximum matching.

7.3.7 Complexity

This algorithm is polynomial as is the previous one of which it is an extension. The problem of the maximum matching in a graph, in particular in a bipartite graph, led to the introduction in the 1960s of the concept of a "good algorithm", that is a polynomial algorithm, the starting point of complexity theory.

Figures 7.3 and 7.4 give an example of the application of the two previous algorithms. The edges of the current matching are in bold. The Hungarian tree with x_3 as its root, found after two augmentations of the matching, enables us, by application of the procedure matching_saturating_X, to see that the bipartite graph considered has no matching saturating $X = \{x_1, \ldots, x_6\}$, because set $\{x_1, x_2, x_3\}$ cannot be matched with set $\{y_2, y_3\}$. By continuing with the procedure `maximum_matching` we obtain, after three more augmentations, a maximum matching.

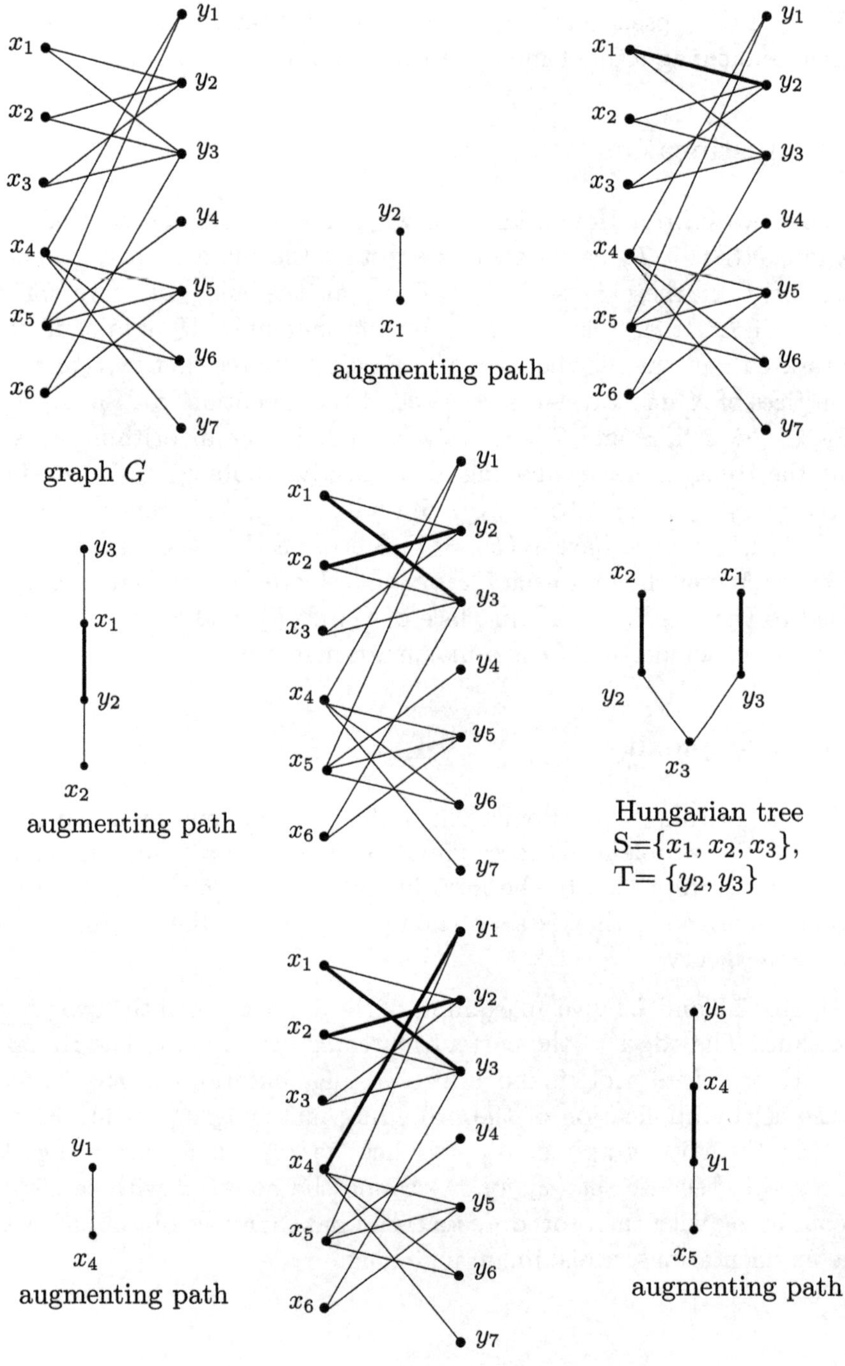

Figure 7.3. *An example of the application of the Hungarian method and of the maximum matching algorithm (continued in Figure 7.4)*

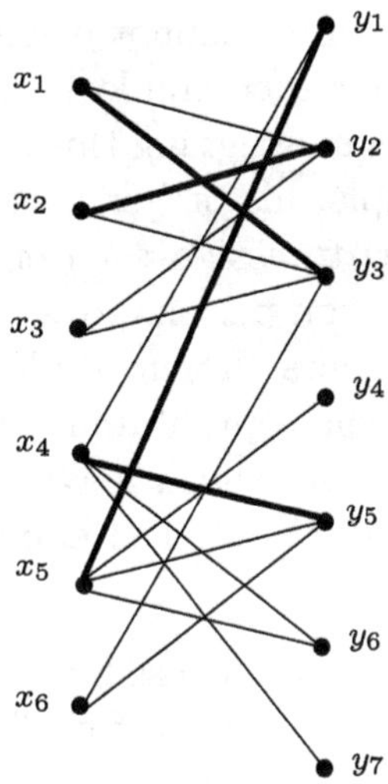

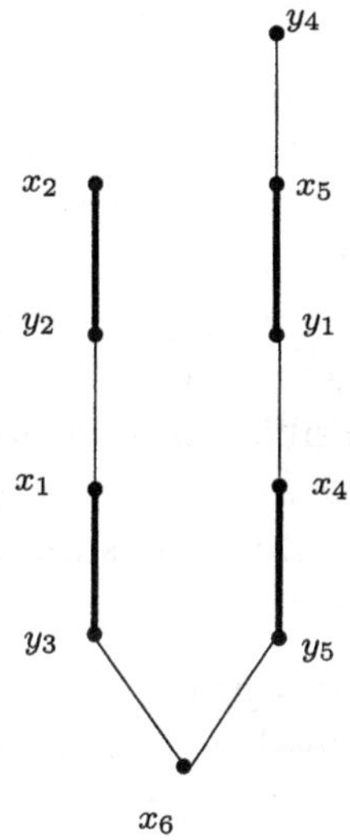

alternating tree with augmenting path

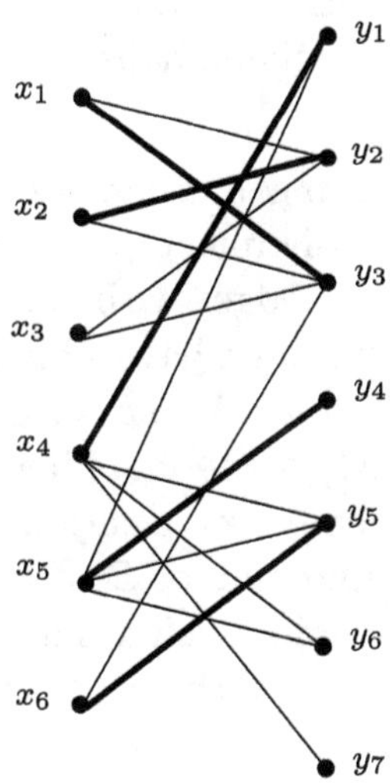

maximum matching

Figure 7.4. *(continuation of Figure 7.3)*

7.4 Optimal assignment problem

Let us again imagine employees to be assigned to positions, but this time instead of only knowing if such employees can or cannot occupy such positions, as in the previous problem, we have for each employee–position pair a value representing the interest that there is in assigning this employee to this position. This value measures, for example, his or her profitability for the position considered. The problem is now to assign each employee, still with one employee per position and one position per employee, so that the sum of the values of the employee-position pairs chosen would be the greatest possible. In other words, looking at it from a profitability point of view, this means finding an assignment for employees which maximizes the total profitability, the sum of the specific profitability of each employee in his or her position.

Note that *a priori* any assignment of the employees is possible. What we are looking for here is an assignment which brings the greatest profitability. The general difficulty is therefore to find an optimal element following a certain measure, an element which we know exists but within a set having such a large number of elements that it is not possible to consider them all within a reasonable timespan. Indeed, observe that if there are n employees and an equal number of positions, the number of possible assignments is $n!$, a quantity which rapidly makes the exhaustive method of looking at each case impractical.

The problem is modeled using a complete bipartite graph defined in the same manner as for the assignment problem in the previous section, but by in addition weighting the edges with the values associated with the employee-position pairs. Therefore, we have a bipartite graph $G = (X, Y, E)$, where X is the set of the employees, Y the set of the positions, with $|X| = |Y|$, and the edges are weighted by the mapping $v : E \to \mathbb{R}$, where $v(xy)$ is the value associated with the employee-position pair $xy \in E$. The problem is therefore to find a perfect matching of G of which the sum of the values of its edges is the greatest possible. Such a matching is said to be *optimal.* The result which we will now give is the key to the algorithmic solution to this problem.

Let a mapping be $f : X \cup Y \to \mathbb{R}$, verifying for any $x \in X$ and $y \in Y$ the condition:

$$(*) \qquad f(x) + f(y) \geq v(xy)$$

and let us define the spanning subgraph H of G induced by the set of the edges xy which verify the equality in (*), that is $f(x) + f(y) = v(xy)$.

LEMMA 7.4. *Any perfect matching of H is an optimal matching of G.*

Proof. For any perfect matching M of G, we have, by applying the condition (*) at each edge of M and by making the sum:

$$v(M) \leq f(X \cup Y)$$

where $v(M) = \sum_{e \in M} v(e)$ and $f(X \cup Y) = \sum_{z \in X \cup Y} f(z)$. Therefore, let M be a matching for which we have the equality:

$$v(M) = f(X \cup Y)$$

This matching is necessarily optimal, because it is of the greatest possible value. Also, when M is a perfect matching of H, therefore also a perfect matching of G, it precisely verifies this equality since it verifies it on each of its edges, by definition of H. □

7.4.1 Kuhn-Munkres algorithm

With lemma 7.4 the path is traced. We have to find a graph such as H, having a perfect matching, that is, an X-saturating matching (since with the hypothesis $|X| = |Y|$, any matching which saturates X also saturates Y). To find such a matching we have what we need with the Hungarian method seen above. However, a graph H remains to be found. Function f is called a *potential function* (since it is defined on the vertices), graph H associated with the *equality graph* associated with function f. The graph H that fits will only be found at the end of the research process. To start, a first potential function is necessary. We can always put, for example:

$$f(z) = \begin{cases} \max\limits_{y \in Y} v(zy) & \text{if } z \in X \\ 0 & \text{if } z \in Y \end{cases}$$

Indeed, it is easy to verify condition (*) for this function. However, there is obviously no reason for the equality graph associated with this function to have in general a perfect matching. Therefore, we have to modify function f, and therefore graph H, until this condition is fulfilled. These modifications will be done every time we come across the non-existence case of an X-saturating matching in H, that is the case of a Hungarian tree. The precise description of these modifications of f is given in the following algorithm which is, in a way, an extension of the matching_saturating_X procedure of section 7.3. Let us specify that graph H is here defined by the neighborhood function denoted by NH, as G was defined above by N, that is $NH(S)$ is the set of the neighbors of the vertices S in H.

The main part of the algorithm, the maximum_val_matching procedure, calls on pseudo-instructions which are made explicit later on.

```
procedure maximum_val_matching(G,v);
begin
  initialize_potential_function(f);
  determine_equality_graph(f,H);
  M:= ∅; I:= X;
  while I ≠ ∅ loop
    r:= a vertex of I;
    look_for_augmenting_path(M,r);
    augment_matching(M,r);
    I:= I \ {r};
  end loop;
end maximum_val_matching;
```

```
initialize_potential_function(f):

  for x ∈ X loop
    f(x):= max(v(xy) | y ∈ Y);
  end loop;
  for y ∈ Y loop
    f(y):= 0;
  end loop;
```

```
determine_equality_graph(f,H):

  for x ∈ X loop NH({x}):= ∅; end loop;
  for y ∈ Y loop NH({y}):= ∅; end loop;
  for x ∈ X loop
    for y ∈ Y loop
      if f(x) + f(y) = v(xy) then
        NH({x}):= NH({x}) ∪ {y};
```

```
        NH({y}):= NH({y}) ∪ {x};
      end if;
    end loop;
  end loop;
```

```
modify_potential_function(f)
(used by look_for_augmenting_path(M,r)):

  m:= min(f(x)+f(y)-v(xy) | x ∈ S, y ∈ Y \ T);
  for x ∈ S loop
    f(x):= f(x) - m;
  end loop;
  for y ∈ T loop
    f(y):= f(y) + m;
  end loop;
```

```
look_for_augmenting_path(M,r):

  S:= {r}; T:= ∅;
  s:= r;
  loop
    if NH(S) = T then
      modify_potential_function(f);
      determine_equality_graph(H);
    end if;
    t:= a vertex of NH(S) \ T;
    parent(t):= a vertex of S ∩ NH({t});
    exit when t M-insatured;
    -- t is M-saturated
    s:= the vertex M-matched to t;
    parent(s):= t;
    S:= S ∪ {s}; T:= T ∪ {t};
  end loop;
```

```
augment_matching(M,r):

  s:= parent(t); M:= M ∪ {st};
  while s ≠ r loop
    t:= parent(s); M:= M \ {st};
    s:= parent(t); M:= M ∪ {st};
  end loop;
```

Note that the pseudo-instruction `augment_matching` is identical to the one given earlier for the maximum matching algorithm. Vertex t is initially the one stemming from the application of `look_for_augmenting_path`.

7.4.2 Justification

We are going to show that the main loop of the `maximum_val_matching` procedure ends after a finite number of iterations with $I = \emptyset$, that is with a matching M of H which saturates X. This matching is a perfect matching of an equality graph, therefore, according to lemma 7.4, it is an optimal matching of G.

Let us consider the case of a vertex $r \in I$ from which no augmenting path may be found. We have then in the current equality graph H a Hungarian tree with root r and with sets of vertices S and T such that $NH(S) = T$. Observe that minimum m, calculated in `modify_potential_function`, is $>$ 0; indeed, an edge xy such that $x \in S$ and $y \in Y \setminus T$ is not in equality graph H (otherwise we would have $y \in NH(S) \setminus T$ and thus $NH(S) \neq T$). Therefore we have:

$$f(x) + f(y) > v(xy)$$

Let $x_0 \in S$ and $y_0 \in Y \setminus T$ for which minimum m is attained, that is, such that:

$$f(x_0) + f(y_0) - v(x_0y_0) = m$$

Let us specify that since this is a Hungarian case, we have $|T| = |S| - 1$, and since $|Y| = |X|$, set $Y \setminus T$ is not empty. Let us put $f_1 = f$, the current function f at that point and f_2 the function modified after (as indicated in `modify_potential_function`). We have (thanks to the previous equality applied to f_1):

$$f_2(x_0) + f_2(y_0) - v(x_0y_0) = f_1(x_0) - m + f_1(y_0) - v(x_0y_0) = 0$$

Let us denote by H_1 the previous equality graph H, associated with f_1, and by H_2 the new one, associated with f_2. The previous equalities indicate that edge x_0y_0, which does not belong to H_1, belongs to H_2. It is easy to verify (modifications are made for that) that graph H_2 contains, as H_1, the alternating tree built. However, this time this tree is no longer a Hungarian case, since, because of the presence of the new edge x_0y_0, there is no longer the equality $NH(S) = T$. This makes it possible to continue looking for an alternating path. Thus each iteration of the **while** loop of the procedure exits from a possible Hungarian case and ends up saturating an extra vertex of X.

This algorithm works because it is always possible to exit from a Hungarian case by modifying the equality graph with the potential function.

NOTE. Function f may be used as a "certificate" to validate the optimal matching found. Apart from the calculation done to find the matching, it is enough to verify the equality $v(M) = f(X \cup Y)$ to be sure that matching M is optimal according to the reasoning held to prove lemma 7.4.

7.4.3 Complexity

As in the Hungarian method, this algorithm has a polynomial complexity which can be evaluated first in $O(n^4)$, where n is the common cardinality of sets X and Y. Indeed, the new operations with regard to the algorithm searching for an X-saturating matching concern the potential function and the equality graph, and are easy to analyze in at worst $O(n^2)$. (Observe that these operations appear in the loops.)

Figure 7.5 gives an example of an application of this algorithm. As it is most often the case in practice, the complete weighted bipartite graph is given by a square matrix $n \times n$ of which the entry (i, j) is the value of edge $x_i y_j$. In the figure we have the succession of potential functions f_0, f_1, f_2 of the corresponding equality graphs H_0, H_1, H_2, and the alternating trees built in each case with, when it happens (case of a Hungarian tree), the calculation of minimum m.

NOTES. 1) As was seen in the previous example, this algorithm means finding in a square table $n \times n$ what we call a *diagonal*, that is n elements such that there is only one per line and only one per column. It is said to be a *maximal* diagonal when the sum of its elements is the greatest possible. It is easy to see that such a diagonal corresponds to an optimal matching of the complete bipartite graph represented by the matrix.

In the example given in Figure 7.5 such a diagonal is found with the entries $(1, 2), (2, 5), (3, 4), (4, 3), (5, 1)$.

	y_1	y_2	y_3	y_4	y_5	f_0	f_1	f_2
x_1	1	2	-1	0	0	2	2	2
x_2	2	-1	0	-1	1	2	1	0
x_3	-1	1	-1	2	3	3	3	2
x_4	1	-2	1	-1	0	1	1	0
x_5	2	-1	1	-3	1	2	1	0
f_0	0	0	0	0	0			
f_1	1	0	0	0	0			
f_2	2	0	1	0	1			

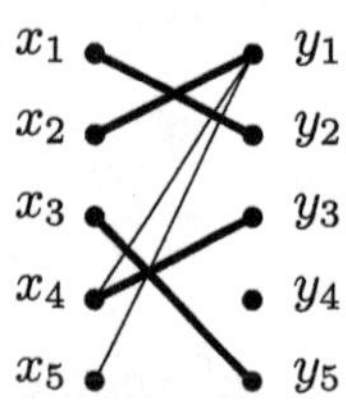

H_0

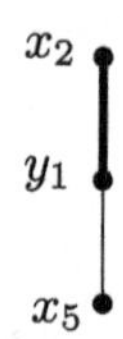

Hungarian tree

	y_2	y_3	y_4	y_5
x_2	3	2	3	1
x_5	3	1	5	1

$m_1 = 1$

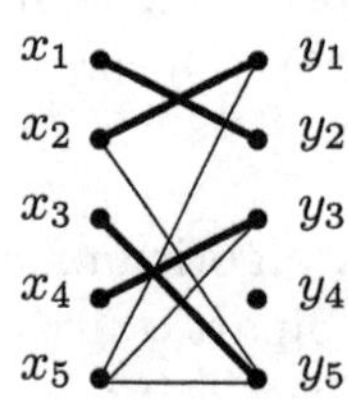

H_1

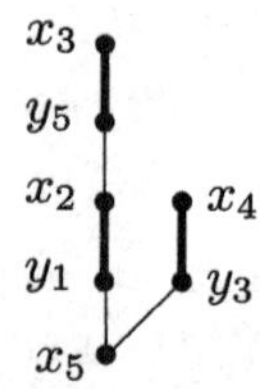

Hungarian tree

	y_2	y_4
x_2	2	2
x_3	2	1
x_4	3	2
x_5	2	4

$m_2 = 1$

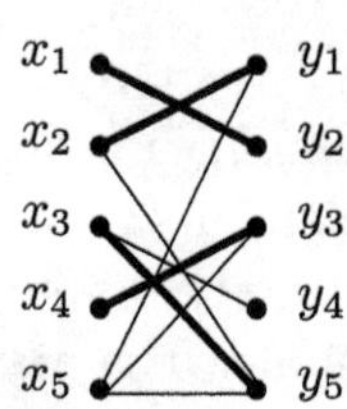

H_2

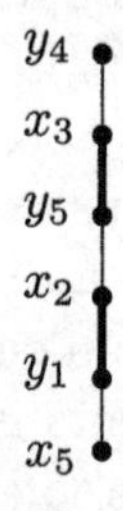

alternating path

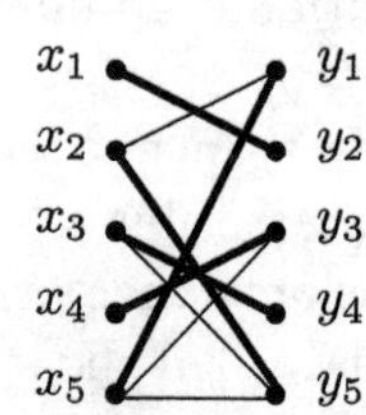

optimal matching

Figure 7.5. *An example of application of the optimal matching algorithm*

2) Looking for a *minimal* diagonal can be done with the same algorithm, which only needs to be applied to the opposite matrix (all entries multiplied by -1). We thus solve the problem of the *optimal* matching in the sense of *minimum* matching.

7.5 Exercises

7.1. Show that the k-cube (defined in Chapter 2) has a perfect matching.

+7.2. a) Show that Hall's theorem (theorem 7.2) can be expressed in the following more general manner: given a bipartite graph $G = (X, Y, E)$, we have $\nu(G) = |X| - \text{def}(X)$, where $\text{def}(X) = \max_{S \subseteq X}(|S| - |N(S)|)$ ("def" for *deficiency*). Note that $\text{def}(X)$ is an integer ≥ 0 because in particular $\text{def}(\emptyset) = 0$.

b) Show the equivalence of the three theorems: Hall, generalized Hall (established in the previous question) and König (theorem 7.3), by showing that generalized Hall implies König which implies Hall.

7.3. Show that a tree T has at most one perfect matching, and that it actually has one if and only if it verifies $p_i(T - x) = 1$ for any vertex x of T, where $p_i(T - x)$ is the number of connected components of $T - x$ which have an odd number of vertices (*the necessary condition is easy*).

7.4. A famous theorem on matchings, Tutte's theorem, states that a graph $G = (X, E)$ has a perfect matching if and only if it verifies $p_i(G-U) \leq |U|$ for any $U \subset X$, where $p_i(G - U)$ is the number of connected components of $G - U$ having an odd number of vertices. Show the necessary condition (*the sufficient condition is harder to demonstrate*).

7.5. Find a *minimal* diagonal in the matrix in Figure 7.5.

*7.6. Five houses, $M_1, \ldots, M_5$, are for sale. The price asked for house M_i is p_i $(i = 1, \ldots, 5)$. These prices are given in ten thousands of euros in the following table:

Houses	M_1	M_2	M_3	M_4	M_5
Prices	7	12	10	15	9

There are five potential buyers, designated by $A_1, \ldots, A_5$. Buyer A_i estimates the price of house M_i to be q_{ij}. The q_{ij} are given in ten thousands of euros in the following table:

	M_1	M_2	M_3	M_4	M_5
A_1	8	14	9	15	7
A_2	9	11	10	14	8
A_3	6	13	9	17	10
A_4	8	10	11	14	7
A_5	9	11	11	12	8

a) Show that it is impossible for the five houses be sold to the five buyers, one house per buyer in such a manner that the buyer buys a house for which the asking price is lower or equal to the price which he had estimated for it (that is if A_i buys M_j, we must have $p_j \leq q_{ij}$). Show that this becomes possible if the asking price for house M_5 is brought down from 9 to 7.

b) We keep for this question the new price of M_5, that is 7, the other prices remaining the same. Let us call the *estimated gain* of a buyer A_i who buys house M_j the difference between his estimate and the asking price (which he pays), that is the quantity $q_{ij} - p_j$. Find how to assign the purchase of the five houses to the buyers so that the sum of the estimated gains by the buyers will be the greatest possible.

Chapter 8

Flows

The concept of flow is, like that of matching, another of the major concepts of graph theory. It has an interesting algebraic aspect and has many major applications, such as transportation problems.

8.1 Flows in transportation networks

A *transportation network* R is defined as: a strict connected digraph $G = (X, A)$ with two disjoint sets, $S \subseteq X$, the set of the *sources* and $T \subseteq X$, the set of the *sinks*, and a weighting of the arcs, $c : A \to \mathbb{N}$ (natural numbers), called the *capacity*. We will also allow the (positive) value ∞ as the capacity of an arc.

Let us specify a few points of terminology and notation. For $a \in A$, $c(a)$ is the *capacity* of arc a. The vertices of $X \setminus (S \cup T)$ are called *intermediate vertices*. For $Z \subseteq X$ we denote as $\omega^+(Z)$ the set of the arcs of G of the form (x, y) with $x \in Z$ and $y \notin Z$, and $\omega^-(Z)$ the set of the arcs of the form (x, y) with $x \notin Z$ and $y \in Z$. The arcs of $\omega^+(Z)$ are the *arcs exiting from* Z and those of $\omega^-(Z)$ are the arcs *entering into* Z. When $Z = \{x\}$, we talk of arcs entering into x and exiting from x.

A *flow* in a transportation network R is a mapping $f : A \to \mathbb{N}$ verifying:

1) $0 \leq f(a) \leq c(a)$ for any $a \in A$,

2) $\sum_{a \in \omega^+(\{x\})} f(a) = \sum_{a \in \omega^-(\{x\})} f(a)$ for any $x \in X \setminus (S \cup T)$.

This second condition is called the *law of flow conservation* for each intermediate vertex, also known in electricity as *Kirchhoff's law.*

For the general case $Z \subseteq X$, $f^+(Z) = \sum_{a \in \omega^+(Z)} f(a)$ and $f^-(Z) = \sum_{a \in \omega^-(Z)} f(a)$. Let us specify that if $\omega^+(Z) = \emptyset$ then $f^+(Z) = 0$, likewise for $\omega^-(Z)$ and $f^-(Z)$. Furthermore, in the case $Z = \{x\}$ we will write $f^+(x)$ instead of $f^+(\{x\})$, and likewise for $f^-(x)$. Condition 2 can also be written as:

2') $f^+(x) = f^-(x)$ for any $x \in X \setminus (S \cup T)$.

We add the following condition which assigns the vertices of S as "sources" and the vertices of T as "sinks" with regard to the flow:

3) $f^+(s) - f^-(s) \geq 0$ for any $s \in S$ and $f^-(t) - f^+(t) \geq 0$ for any $t \in T$.

Figure 8.1 gives an example of a flow in a transportation network.

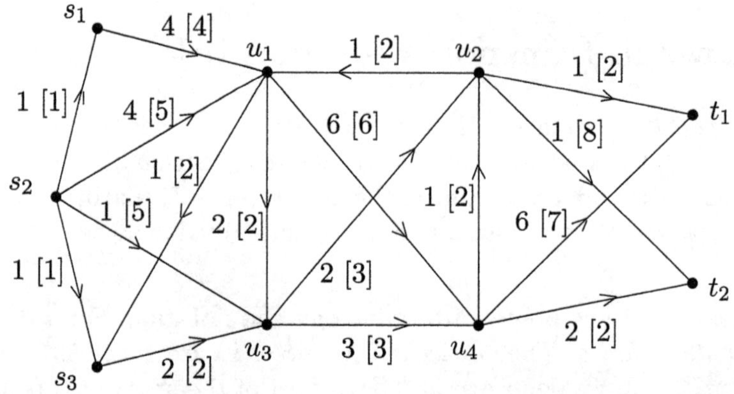

Figure 8.1. *In this representation of a transportation network, the capacities of the arcs are given between brackets. We have here* $S = \{s_1, s_2, s_3\}$ *and* $T = \{t_1, t_2\}$. *A flow is given by the values indicated to the left of the capacities of each arc. These representation conventions also apply to the figures which follow*

NOTE. In any network, there is always what is called the *zero flow*, meaning the flow taking value 0 on each arc.

8.1.1 Interpretation

A network such as the one in Figure 8.1 can be interpreted as, for example, a network for transporting goods or a water transportation system. Sources s_1, s_2, and s_3 are then production or supply centres, from which the flow starts (and to which it may return). Sinks t_1 and t_2 are centers where some flow is consumed (it may also start again from there).

The following result, very intuitive but not completely self-evident to prove, introduces the concept of the value of flow transiting in the network.

LEMMA 8.1. *We have, using the previous notations:*

$$f^+(S) - f^-(S) = f^-(T) - f^+(T)$$

Proof. We can write successively:

$$\begin{aligned}
0 &= \sum_{x \in X} \left(f^+(x) - f^-(x)\right) \\
&= \sum_{x \in S} \left(f^+(x) - f^-(x)\right) + \sum_{x \in X \setminus (S \cup T)} \left(f^+(x) - f^-(x)\right) \\
&\quad + \sum_{x \in T} \left(f^+(x) - f^-(x)\right) \\
&= \sum_{x \in S} \left(f^+(x) - f^-(x)\right) + \sum_{x \in T} \left(f^+(x) - f^-(x)\right) \\
&= \left(f^+(S) - f^-(S)\right) + \left(f^+(T) - f^-(T)\right)
\end{aligned}$$

The sum over $X \setminus (S \cup T)$ is zero according to condition 2' of the definition of flows. The equality $\sum_{x \in S}(f^+(x) - f^-(x)) = f^+(S) - f^-(S)$ of course has to be verified (which is easy) as well as the equivalent one for T. □

We can verify in the example in Figure 8.1 that we really have this equality.

Let us call *net flow out of* $Z \subset X$ the quantity $f^+(Z) - f^-(Z)$, and *net flow into* Z the quantity $f^-(Z) - f^+(Z)$. Lemma 8.1 states that the net flow out of S is equal to the net flow into T. This result is obviously based on the property of conservation of the flow at the intermediate vertices. This common value $f^+(S) - f^-(S) = f^-(T) - f^+(T)$ is called the *value* of flow f, and denoted as $v(f)$. The problem dealt with in this chapter is that of *maximum flow*, that is, of finding in a network a flow with the greatest possible value.

8.1.2 Single-source single-sink networks

Assume that $|S| = |T| = 1$. We will denote by s the single source and by t the single sink of the network. In fact, we can, without losing generality, always suppose that this is the case. In addition, it is also possible to impose the condition that there are no arcs entering into s, nor exiting from t, so that s is a single source vertex and t a single sink vertex in the digraph of the network. Let us justify this.

Let R be a network such as that defined above. Let us put $S = \{s_1, s_2, \ldots, s_p\}$ and $T = \{t_1, t_2, \ldots, t_q\}$. Let us add to this network a vertex s with an arc (s, s_i) of capacity ∞ for each of the vertices s_i $(i = 1, 2, \ldots, p)$, and a vertex t with an arc (t_j, t) of capacity ∞ for each of the vertices t_j $(j = 1, 2, \ldots, q)$. Let us denote by R' the network thus obtained with sets of sources and of sinks $S' = \{s\}$ and $T' = \{t\}$. Therefore it is a single-source single-sink network. It is possible to verify that any flow of R corresponds to a flow of R' of the same value and vice versa. Indeed, if f is a flow of R, let us define f' in the following way:

$$f'((s, s_i)) = f^+(s_i) - f^-(s_i) \quad \text{for } i = 1, 2, \ldots, p$$

and

$$f'((t_j, t)) = f^-(t_j) - f^+(t_j) \quad \text{for } j = 1, 2, \ldots, q$$

Thus, the vertices of S and T in R become in R' intermediate vertices, where the flow conservation law is respected. For all the other arcs of R' the value of f' is taken to be equal to that of f. It is easy to verify that a flow f' of R' is thus truly defined, and that this flow has an equal value to that of f. Note in particular that the values of f' are really ≥ 0 on the arcs added to R as a result of condition 3 of the definition of flows.

Conversely, given a flow R', we can directly deduce from it a flow in R, of the same value, by considering its restriction to the arcs of R.

In the case of a flow in a network with a single source s and a single sink t, the value of flow f is $v(f) = f^+(s) = f^-(t)$ because, since there are no arcs entering into s nor exiting from t, we have $f^-(s) = 0$ and $f^+(t) = 0$.

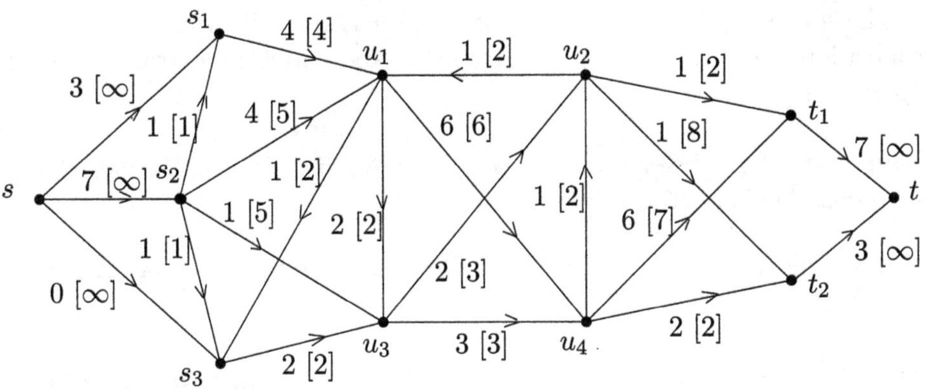

Figure 8.2. *Network of Figure 8.1 modified to become a single-source single-sink network*

Figure 8.2 gives network R' corresponding to network R of Figure 8.1.

NOTE. By adding to the network an arc (t, s), of capacity ∞, we obtain a network in which the law of conservation of the flow applies at any vertex.

8.2 The max-flow min-cut theorem

We now suppose that the networks considered are single-source, single-sink networks. The concept of cut, dual to that of flow, is, as we will see, essential in this study. A *cut* of network R is a set of arcs of the network of the form $K = \omega^+(Z)$, where $Z \subset X$ such that $s \in Z$ and $t \notin Z$. The *capacity* of a cut K is the sum of the capacities of its arcs, that is $c(K) = \sum_{a \in K} c(a)$. The following simple proposition plays an essential role.

PROPOSITION 8.1. *For any flow f and any cut K, we have:*

$$v(f) \leq c(K)$$

Proof. We have successively:

$$c(K) \geq f^+(Z) \geq f^+(Z) - f^-(Z) = \sum_{x \in Z} (f^+(x) - f^-(x)) = f^+(s) = v(f)$$

(Note that $f^-(Z) \geq 0$, which justifies the second inequality.) □

A *minimum cut* is a cut of the lowest possible capacity.

The following fundamental property results directly from the previous proposition (as for maximum matching and minimum transversal in Chapter 7):

COROLLARY 8.1. *Given a flow f and a cut K, if $v(f) = c(K)$ then f is a maximum flow and K a minimum cut.*

It is this last result which will allow us to recognize that a flow is maximum.

NOTE. If we allow the possibility of arcs with infinite capacities, as we have done above to go back to the case of single-source, single-sink networks, there can be cuts with infinite capacities. If all the cuts of a network are in that condition, we have in the network flows with arbitrarily great values, as in the (trivial) example in Figure 8.3. Then, there is, as it were, no maximum flow. When applying the previous corollary, we will suppose implicitly that there is a cut with a finite capacity.

Figure 8.3. *An "unbounded" network*

8.2.1 Concept of unsaturated paths

Let us remember that a *path* in a digraph G is a path of the undirected graph associated with G. This path is defined by a sequence of arcs of the graph which are not necessarily directed in the same direction (which makes it a path and not a *directed* path). Let μ be a path in a network R with a flow f, the path joining source s of the network to a vertex u, and presumed to have a length which is not zero. Let us denote by $(a_1, \ldots, a_k)$ the sequence of its arcs ($k \geq 1$), and let us put:

$$\tau(\mu) = \min_{i=1,\ldots,k} \tau(a_i)$$

where $\tau(a_i) = c(a_i) - f(a_i)$ if arc a_i is directed from s to u in path μ, and $\tau(a_i) = f(a_i)$ if arc a_i is directed from u to s in μ. The path is said to be *unsaturated*, for f, if $\tau(\mu) > 0$.

LEMMA 8.2. *If there is for flow f an unsaturated path joining s to t, then there is a flow f' such that $v(f') > v(f)$.*

Proof. Let μ be an unsaturated path joining s to t. We identify by notation μ with the set of the arcs of this path, and we put μ^+ for the set of arcs of path μ which are directed from s to t, and μ^- for those directed from t to s. Let us define f' as:

$$f'(a) = \begin{cases} f(a) + \tau(\mu) & \text{if } a \in \mu^+ \\ f(a) - \tau(\mu) & \text{if } a \in \mu^- \\ f(a) & \text{if } a \notin \mu \end{cases}$$

It is easy to verify that f' is really a flow of R and that we have the equality $v(f') = v(f) + \tau(\mu)$. Since $\tau(\mu) > 0$, we have $v(f') > v(f)$. □

LEMMA 8.3. *If there is no unsaturated path joining s to t for flow f, then there is a cut K such that $c(K) = v(f)$.*

Proof. Let Z be the set of vertex s and the vertices x such that there is an unsaturated path from s to x. We have by hypothesis $s \in Z$ and $t \notin Z$. For any arc $a = (x, y)$ such that $x \in Z$ and $y \notin Z$ (that is $a \in \omega^+(Z)$) we have $f(a) = c(a)$, otherwise we could extend an unsaturated path from s to x with arc a, which would contradict $y \notin Z$. For any arc $a = (y, x)$ such that $x \in Z$ and $y \notin Z$ (that is $a \in \omega^-(Z)$), we have $f(a) = 0$, otherwise we would have an unsaturated path from s to y contrary to the hypothesis that $y \notin Z$. We can write the following successive equalities (partly as seen above for proving lemma 8.1, note that here $f^-(Z) = 0$):

$$\begin{aligned} c(K) &= f^+(Z) = f^+(Z) - f^-(Z) \\ &= \sum_{x \in Z} \left(f^+(x) - f^-(x)\right) = f^+(s) = v(f) \end{aligned}$$

□

In consideration of these results, we call any unsaturated path joining s to t an *incrementing path* for flow f. We have:

PROPOSITION 8.2. *A flow is maximum if and only if there is no incrementing path for this flow.*

Proof. Directly from lemmas 8.2 and 8.3, and corollary 8.1. □

THEOREM 8.1 (Ford-Fulkerson). *In any network, the value of a maximum flow is equal to the capacity of a minimum cut.*

Proof. Directly from proposition 8.2 and lemma 8.3. □

The diagram at the bottom of Figure 8.6 in the following section will give an example of what we have seen above.

8.3 Maximum flow algorithm

After giving the criterion for a flow to be maximum, it is now important, as in any optimization problem, to be able to really give a maximum flow. This is what the classic Ford-Fulkerson's algorithm, given below, does. It is based on the previous theoretical development. Given a network R and a flow f (it is always possible initially to consider the zero-flow), the crucial point is finding an incrementing path. We know that if such a path does not exist, the flow considered is maximum (proposition 8.2). The algorithm defines, in `look_for_incrementing_path`, how to proceed for such a search. When an incrementing path is found, the statement `augment_flow` applies what is defined in the proof of lemma 8.2. Let us specify some other points.

The digraph of the network is given in the algorithm by the sets of arcs $\omega^+(\{x\})$ and $\omega^-(\{x\})$ defined above ($\omega^+(\{x\})$ is the set of arcs exiting from x and $\omega^-(\{x\})$ the set of arcs entering into x). In `look_for_incrementing_path`, a labeling process of the vertices is implemented with mapping l defined on the vertices and with values in $\mathbb{N} \cup \{\infty\}$, which makes it possible to spot the vertices y linked by an unsaturated path μ from vertex s.

Thanks to the calculation of minima on l, the label of the vertex under consideration, $l(y)$, is the quantity $\tau(\mu)$ defined above. Thus when an incrementing path is found, a path from s to t, the "incrementation" value $\tau(\mu)$ is the $l(t)$ of the algorithm. It is this value which is used again in `augment_flow` to modify the current flow, as in the proof of lemma 8.2. Set L is the set of vertices said to be "labeled", that is those on which mapping l is defined. Set Z is that of the labeled and "explored" vertices, that is those considered for possibly labeling their neighbors in turn (loop **for** in `look_for_incrementing_path`). We still have $Z \subseteq L$. Array `pred` defines the unsaturated path built in `look_for_incrementing_path` in the following manner: for a vertex y of the path, different from s, we have $\text{pred}(y) = (x, a, \epsilon)$ where x is the predecessor of y in the unsaturated path, a is the arc joining x and y, from x to y or from y to x depending on the cases, and ϵ is the integer +1 or -1 following respectively the two preceding cases (pred(y) can be seen in a program as a record having those three components). The integer ϵ makes it possible to know if quantity $l(t)$ given by the incrementing path has to be added to or taken out of flow on arc a in `augment_flow`. Finally, the Boolean `found` indicates whether an incrementing path has been found or not.

```
procedure maximum_flow(R);
begin
  -- initialization
  for a ∈ A loop
    f(a):= 0;
  end loop;
  -- treatment
  loop
    look_for_incrementing_path(f);
    if found then
      augment_flow(f);
    else
      exit;
    end if;
  end loop;
end maximum_flow;
```

The two pseudo-instructions `look_for_incrementing_path` and `augment_flow` are made explicit in what follows, where `L` and `Z` are auxiliary sets of vertices.

```
augment_flow(f):

y:= t;
loop
  (x,a,ε):= pred(y);
  f(a):= f(a) + ε.l(t);
  exit when x = s;
  y:= x;
end loop;
```

```
look_for_incrementing_path(f):

l(s):= ∞; L:= {s}; Z:= ∅;
```

```
loop
  x:= a vertice of L \ Z;
  Z:= Z ∪ {x};
  for a = (x,y) ∈ ω+({x}) loop
    if y ∉ L and f(a) < c(a) then
      l(y):= min(l(x),c(a)-f(a));
      L:= L ∪ {y};
      pred(y):= (x,a,+1);
    end if;
  end loop;
  for a = (y,x) ∈ ω-({x}) loop
    if y ∉ L and f(a) > 0 then
      l(y):= min(l(x),f(a));
      L:= L ∪ {y};
      pred(y):= (x,a,-1);
    end if;
  end loop;
  if t ∈ L then
    found:= true;
    exit;
  end if
  if L \ Z = ∅ then
    found:= false;
    exit;
  end if;
end loop;
```

8.3.1 Justification

There is no real need to justify this algorithm which works just as in the preceding theoretical development: augmentation of the flow as long as an incrementing path may be found, that is, as long as one exists. When there are no more incrementing paths, the flow is maximum (proposition 8.2). We can verify this by application of lemma 8.3 and cut K of which the capacity is equal to the value of the flow.

NOTES. 1) The search for an augmenting path is done by an unsaturated tree growth process from vertex s, just as with alternating trees in the Hungarian method for matchings (Chapter 7).

2) We exit this search when sink t of the network is reached. Indeed, it is useless to continue the search further since we have, in the unsaturated tree, a path from s to t. In the algorithm, to be more specific in `look_for_incrementing_path`, we can exit as early as possible by exiting from the first interior **for** loop as soon as condition $y = t$ is realized (be careful: it is an exit from *two* loops). With the hypothesis given above that vertex t only has entering arcs, it is really the first **for** loop because the only possibility for reaching t is then $a = (x, y) \in \omega^+(\{x\})$. (We will have in this case $y = t$.)

3) Set Z makes it possible to "certify" that the flow is maximum by simply verifying the equality $v(f) = c(K)$, where K is the cut defined by Z. This equality can also be verified by the fact, here characteristic of the optimum, that any arc exiting from Z is saturated for f, that is, verifies $f(a) = c(a)$, and any arc entering into Z has a zero-flow, that is, verifies $f(a) = 0$.

4) In practice, it is possible to reduce the number of iterations of the main loop of the procedure by starting from a non zero-flow. Indeed, the algorithm works from any flow (including a maximum flow, in which case we will stop with the first iteration of the main loop of the procedure).

An example of an application of this algorithm is given in Figures 8.4, 8.5 and 8.6. The vertices labels are given on each vertex concerned in a square, following the progression of the algorithm. This defines the vertices of set L. The vertices of Z are indicated in bold. The arcs of the unsaturated tree built are in bold and those defining an incrementing path are bolder. The first diagram represents the network with zero-flow. The following four diagrams give the details of the first search for an augmenting path leading to the first non zero-flow, with a value equal to 2 (sixth diagram). The next iterations of the search are not detailed: a new augmenting path gives a flow with a value equal to 5, then another gives a flow with a value of 8. This last flow is maximum as is shown by the set of vertices Z, which defines a cut of capacity equal to 8, which is therefore equal to the value of the maximum flow (last diagram). The arcs of the minimum cut are given as dotted lines. Observe that any exiting arc is saturated and any entering arc has a zero-flow, which illustrates the third note made above.

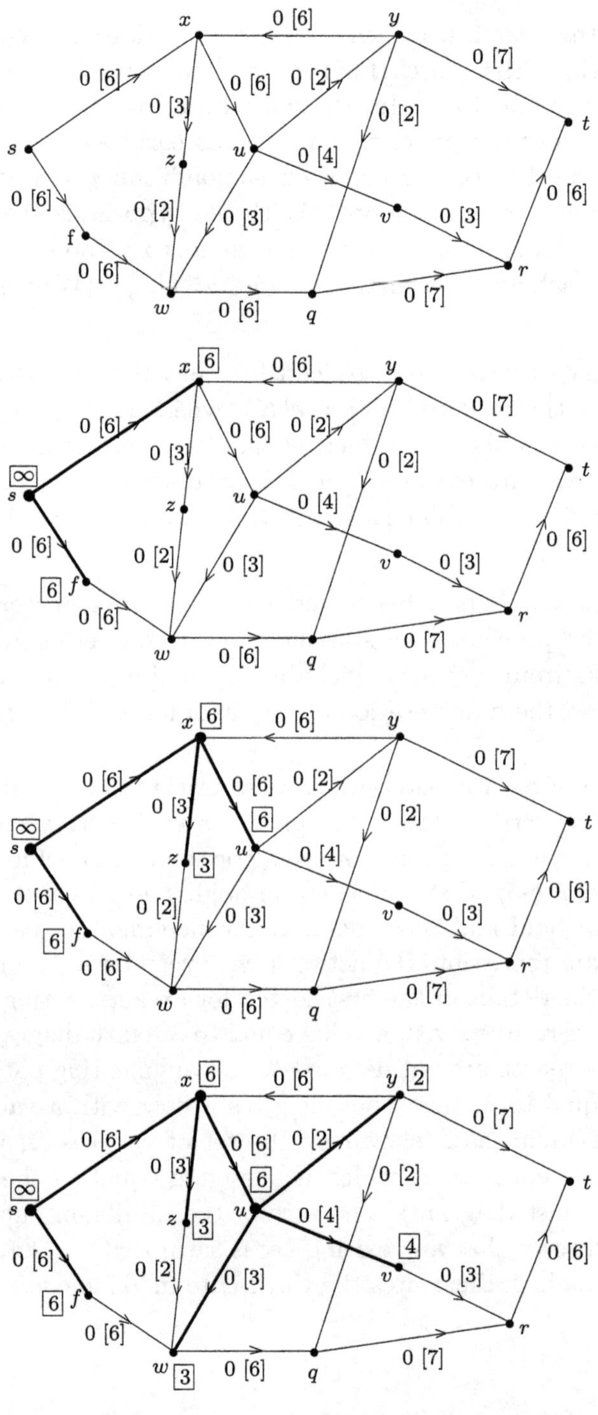

Figure 8.4. *An example of an application of the maximum flow algorithm (to be continued in Figures 8.5 and 8.6)*

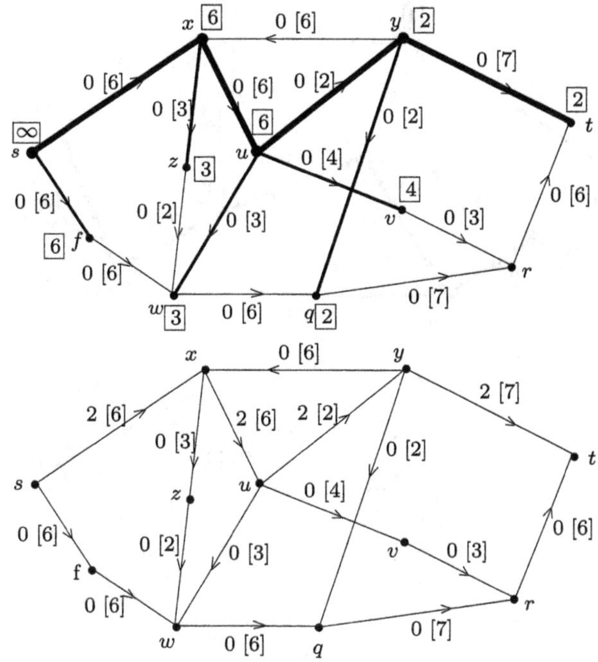

(flow of value 2)

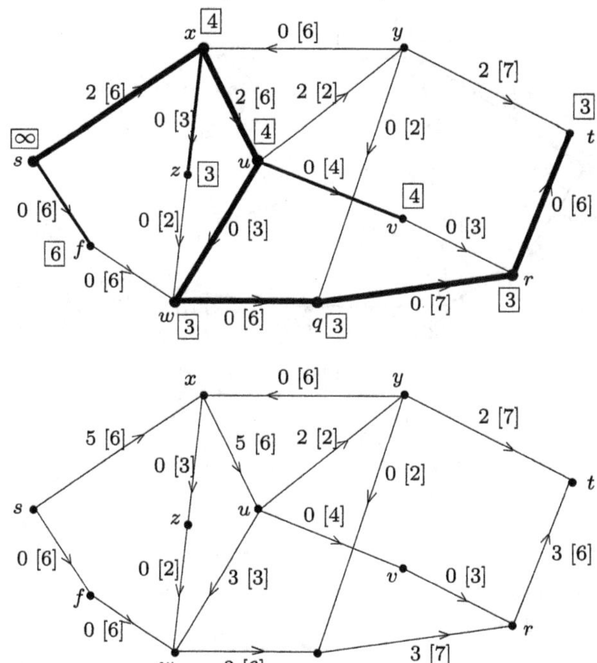

(flow of value 5)

Figure 8.5. *(continuation of Figure 8.4)*

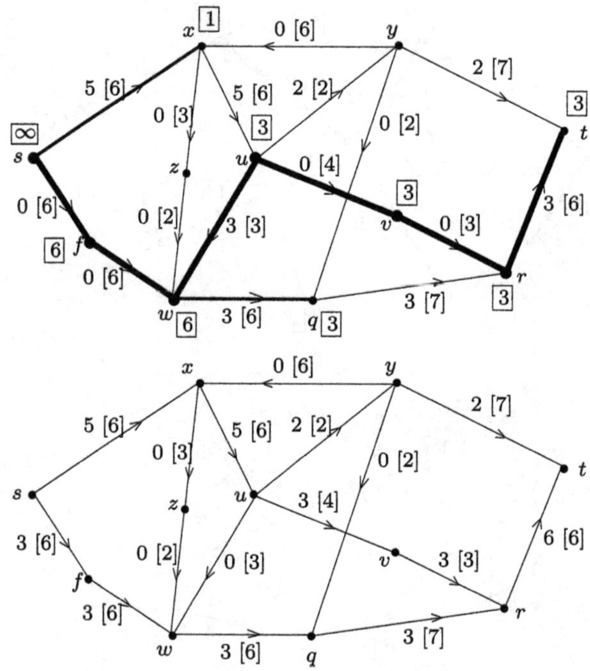

(flow of value 8)

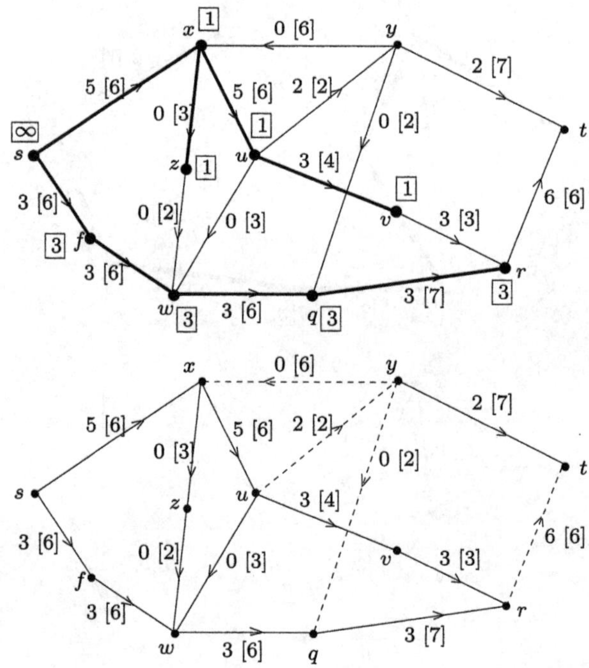

(cut of capacity 8)

Figure 8.6. *(continuation of Figure 8.5)*

8.3.2 Complexity

Unexpectedly, this algorithm, as it is, is not of polynomial complexity. Indeed, consider the network in Figure 8.7, in which capacity q is an integer > 0. Let us suppose that we apply the algorithm, starting with the zero-flow, by increasing the flow first by the augmenting path (s, x, y, v, w, t), then by the augmenting path (s, u, v, y, z, t), and so on, alternating between these two paths. Each augmentation is only of 1 (since the capacity of arc (y, v) equals 1), and so at least $2q$ iterations of the main loop of the algorithm will be necessary to reach the maximum flow, which is easy to find and which † has a value equal to $2q$.

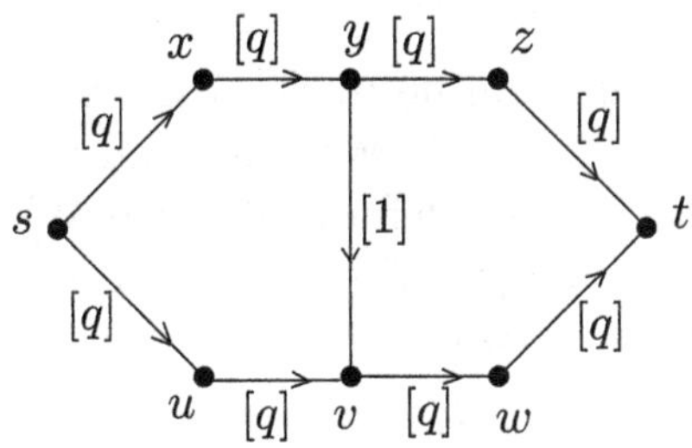

Figure 8.7. *A case of non-polynomial complexity*

The general problem which is seen through this example is that the number of iterations of the main loop of the algorithm may not be bounded with regard to the size of the network defined by vertices number n and arcs number m of the digraph. Indeed, this number of iterations may be increased following the value of the capacities of the arcs, values which are independent from n and m. However, there is a simple remedy to this problem. If we look closely at the previous example, we will see that the cause of the number of iterations which is of the order of integer q results from the fact that we have gone back over labeled vertices to explore them in an order which was not the one in which they had been labeled, which is not forbidden (let us remember that a vertex is said to have been "explored" when its neighbors have been considered to be possibly labeled in turn). Respecting the rule: *explore the vertices in the order in which they have been labeled,* we see in the previous example that we reach a maximum flow in only two iterations! This remedy is general and makes it possible to show that, thus completed, the maximum flow algorithm is polynomial and with a complexity which may already be evaluated as $O(mn)$.

The practical implementation in the algorithm of this rule, which may be summed up as "first labeled, first explored", is done in `look_for_incrementing_path` by storing the vertices of $L \setminus Z$ in a queue: a vertex x taken from $L \setminus Z$ is then removed from the queue (at the front) and a vertex y inserted in L is inserted into the queue (at the rear). This method corresponds to searching for an incrementing path with a *breadth-first search* (see Chapter 6) of the network digraph. This means searching for augmenting paths with the shortest possible lengths (length meaning here the number of arcs).

8.4 Flow with stocks and demands

Let us come back to a general network with sets of sources and sinks S and T which are of cardinality ≥ 1. We are given the following additional data: a mapping $\sigma : S \to \mathbb{N}$ which defines what we call *stocks* at the sources of the network, and a mapping $\tau : T \to \mathbb{N}$ which defines what we call the *demands* at the sinks of the network.

In such a network, a flow is called *feasible* if it satisfies the following conditions:

$$f^+(s) - f^-(s) \leq \sigma(s) \quad \text{for any } s \in S$$

$$f^-(t) - f^+(t) \geq \tau(t) \quad \text{for any } t \in T$$

The interpretation of these inequalities in terms of stocks and demands, following the terminology, is clear: a feasible flow must carry through the network a flow which does not exceed the stocks available at the sources and which covers the demands at the sinks.

There is no reason why a feasible flow should always exist. For example there is none when the sum of the demands at the sinks, $\tau(T) = \sum_{t \in T} \tau(t)$, exceeds the sum of the stocks at the entries, $\sigma(S) = \sum_{s \in S} \sigma(s)$. A necessary condition of existence for a feasible flow is therefore $\sigma(S) \geq \tau(T)$. The following classic result gives a general necessary and sufficient condition for the existence of a feasible flow.

THEOREM 8.2 (Gale). *A network R, of which X is the set of vertices, with stocks for S defined by σ and with demands on T defined by τ, allows a feasible flow if and only if we have for any $U \subseteq X$:*

$$c(\omega^-(U)) \geq \tau(T \cap U) - \sigma(S \cap U)$$

Proof. This condition naturally expresses that for $U \subseteq X$, the entry capacity in U is greater than or equal to the demand in U minus the stock in U. It can be verified directly that this condition is necessary for a flow f:

$$\begin{aligned} c\big(\omega^-(U)\big) &\geq f^-(U) \geq f^-(U) - f^+(U) \\ &= \sum_{x \in U} \big(f^-(x) - f^+(x)\big) \\ &= \sum_{x \in S \cap U} \big(f^-(x) - f^+(x)\big) + \sum_{x \in T \cap U} \big(f^-(x) - f^+(x)\big) \\ &\geq -\sigma(S \cap U) + \tau(T \cap U) \end{aligned}$$

because on the one hand:

$$\sum_{x \in S \cap U} \big(f^+(x) - f^-(x)\big) \leq \sum_{x \in S \cap U} \sigma(x) = \sigma(S \cap U)$$

and on the other:

$$\sum_{x \in T \cap U} \big(f^-(x) - f^+(x)\big) \geq \sum_{x \in T \cap U} \tau(x) = \tau(T \cap U)$$

(the calculations made here are in part analogous to those made for the proof of lemma 8.1).

To show the sufficient condition, let us associate with network R network R' with a single source s and a single sink t built from R as indicated in section 8.1.2, except for the capacities of the added arcs. Let us put for network R, $S = \{s_1, s_2, \ldots, s_p\}$ and $T = \{t_1, t_2, \ldots, t_q\}$. For each arc (s, s_i) of R', let us put its capacity $c'(s, s_i)$ equal to $\sigma(s_i)$ $(i = 1, 2, \ldots, p)$, and for each arc (t_j, t) of R', let us put its capacity $c'(t_j, t)$ equal to $\tau(t_j)$ $(j = 1, 2, \ldots, q)$. It is easy to see that there is a feasible flow in R as soon as there is a flow in R' which *saturates* the "sink" arcs (t_j, t) (flow equal to the capacity of these arcs). However, according to theorem 8.1 applied to R', such a flow exists if any cut of R' has a capacity greater than or equal to $\sum_{j=1}^{q} \tau(t_i) = \tau(T)$, because a maximum flow will then have a value $\geq \tau(T)$ (in fact equal to $\tau(T)$). Let K' be a cut of R' defined by the set of vertices Z', with $s \in Z'$ and $t \notin Z'$. Let us put $U = X \setminus Z'$. We have the following, observing in particular that $t \notin U$ since $t \notin X$ (here ω^- relates to R):

$$c'(K') = \sigma(S \cap U) + c\big(\omega^-(U)\big) + \tau(T \cap Z')$$

and the condition $c'(K') \geq \tau(T)$ then becomes:

$$\sigma(S \cap U) + c\big(\omega^-(U)\big) + \tau(T \cap Z') \geq \tau(T)$$

that is, taking into account $\tau(T) - \tau(T \cap Z') = \tau(T \cap U)$:

$$c\big(\omega^-(U)\big) \geq \tau(T \cap U) - \sigma(S \cap U)$$

which is precisely the condition of the theorem. □

NOTES. 1) With $U = X$, we have the necessary condition previously described, $\sigma(S) \geq \tau(T)$.

2) Another intuitive condition of the existence of a feasible flow is that, in a network considered without stocks and demands, the value of a maximum flow must not be less than the total demand. Let us consider a single-source, single-sink network R'' obtained from network R following the construction given in 8.1.2 (R'' is defined as network R' in the proof of the previous theorem except for the added arcs (s, s_i) and (t_j, t), which have here an infinite capacity). A cut $K'' = \omega^+(Z'')$ in R'' has an infinite capacity as soon as it contains one of the added arcs. The cuts with finite capacities are those defined by a set Z'' such that $Z'' \supseteq S$ and $Z'' \cap T = \emptyset$ (in addition $s \in Z''$ and $t \notin Z''$). Let us suppose that Z'' defines a minimum cut and let us put $U = X \setminus Z''$, with then $U \cap S = \emptyset$ and $U \supseteq T$. The condition of theorem 8.2 makes it possible to write:

$$c(K'') = c\big(\omega^-(U)\big) \geq \tau(T \cap U) - \sigma(S \cap U) = \tau(T)$$

which gives, with the equality $v(f) = c(K'')$ when f is a maximum flow, the indicated condition, that is, that the value of a maximum flow must be greater than or equal to the total demand.

3) It is also interesting to consider the case $U = \{t_j\}$ which gives:

$$c\big(\omega^-(\{t_j\})\big) \geq \tau(\{t_j\}) - \sigma(\emptyset) = \tau(\{t_j\})$$

and shows that the capacity entering into vertex t_j must be at least equal to the demand in t_j, which is clear. It is again interesting to consider the case $U = X \setminus \{s_i\}$ which gives:

$$c\big(\omega^+(\{s_i\})\big) \geq \tau(T) - \sigma\big(S \setminus \{s_i\}\big)$$

thus:

$$\sigma(s_i) - c(\omega^+(\{s_i\})) \leq \sigma(S) - \tau(T)$$

a condition which the reader is left to interpret. †

8.5 Revisiting theorems

The theory of flows in a network, of which we have just developed the bases, is very powerful. It makes it possible, for example, to prove some important theorems as we are going to see with three examples.

8.5.1 Menger's theorem

Let us recall this theorem in its "vertex statement" (Chapter 2, theorem 2.2): *A (undirected) graph G such that $n \geq k+1$ is k-connected if and only if any two distinct vertices of G are joined by k internally vertex-disjoint paths (that is pairwise with no other common vertices other than their ends).*

The part which is the least easy to prove is the necessary condition. It will easily result from the following lemma which expresses the equality of the value of a maximum flow and of the capacity of a minimum cut in a particular network. This result is in fact the original version of the theorem given by Menger, sometimes called the "local" version of the theorem.

LEMMA 8.4. *Let G be a simple graph and let x and y be two non-neighboring vertices of G. The greatest number of pairwise vertex-disjoint paths linking x and y in G is equal to the lowest number of elements of a set of vertices A which separates x and y, that is such that x and y are in different connected components of $G - A$.*

Proof. Let us build a network from graph G in the following way: each edge of G is first replaced by two opposite arcs, giving a digraph, denoted by G', which has the same set of vertices as G. Then, each vertex $z \neq x, y$ of G' is split into two vertices z' and z'' with an arc (z', z'') of capacity 1. Any arc entering into z is replaced by an arc entering into z' with a capacity ∞. Any arc exiting from z is replaced by an arc exiting from z'', also with a capacity ∞. Vertices x and y respectively play the role of source and sink in the network thus defined, which we denote by R (see an example in Figure 8.8). It is not difficult to verify what follows:

- A flow of R defines a set of pairwise vertex-disjoint paths of G (the arcs of the form (z', z'') for which the flow takes the value 1 define the corresponding vertices of the paths of G).

– A finite cut of R defines in G a set of vertices A such that x and y are in different connected components of $G - A$. Indeed, a finite cut of R is necessarily composed of arcs of the form (z', z''), which are the only arcs with finite capacities and which correspond to vertices of G. Ford-Fulkerson's theorem applied to network R then gives the result directly. □

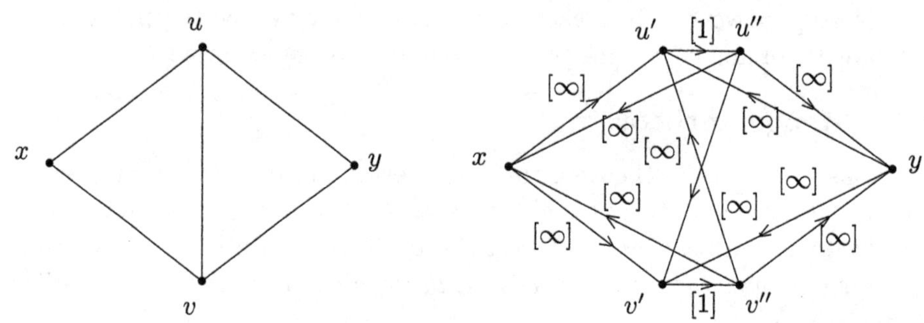

graph G associated network R

Figure 8.8. *Proof of lemma 8.4*

Let us apply lemma 8.4 to a proof of the necessary condition of Menger's theorem. Let G be a k-connected graph, let x and y be two vertices of G. (We may suppose $k \geq 1$, the case $k = 0$ being trivial.) If x and y are not neighbors in G, lemma 8.4 gives the result directly. Indeed, the lowest number of elements of a set A, as stated in lemma 8.4, is greater than or equal to the connectedness number of G, $\kappa(G)$, and G being k-connected we have $\kappa(G) \geq k$. Thus there are a number of vertex-disjoint paths joining x and y at least equal to k. If x and y are neighbors in G, endvertices of an edge e, we can consider graph $G - e$, which we know to be $(k-1)$-connected (see Chapter 2, exercise 2.16). The application of lemma 8.4 then gives $(k-1)$ paths joining x and y, which, with the path defined by edge xy and its endvertices, constitute the required number of k paths.

Let us note that we can similarly prove the "edge" version of Menger's theorem.

8.5.2 Hall's theorem

Let us recall this classic maximum matching theorem in a bipartite graph (Chapter 7, theorem 7.2): *a bipartite graph* $G = (X, Y, E)$ *allows a matching which saturates* X *if and only if for any* $S \subseteq X$ *we have* $|N(S)| \geq |S|$, where $N(S)$ designates the set of the neighbors of the vertices of S. The part which is not obvious, the sufficient condition, is an easy consequence of theorem 8.2. We call the necessary and sufficient condition given in theorem 8.2 *Gale's condition.*

Let us associate with bipartite graph $G = (X, Y, E)$ the transportation network R with stocks and demands, built in the following way. Each edge of G is directed from Y to X and receives a capacity ∞. Set Y is defined as set S of the sources of R, and X as set T of the sinks of R. We define a stock equal to 1 for each source and a demand equal to 1 on each sink. It is easy to see that a feasible flow of R corresponds to a matching which saturates X. The matching is defined by the arcs of R with a flow equal to 1, and, since the demand is satisfied, each vertex of G is in fact saturated by this matching.

It is enough therefore to verify that Hall's condition in G leads to Gale's condition in R (a figure may be useful for following the reasoning below). Let there be $U \subseteq X \cup Y$ (note that the set of the vertices of the graph here is $X \cup Y$, and not simply X as in Gale's theorem). If we don't have $N(U \cap X) \subseteq U \cap Y$, then $\omega^-(U) \neq \emptyset$, $c(\omega^-(U)) = \infty$, and Gale's condition is trivially verified. If $N(U \cap X) \subseteq U \cap Y$, we have $\omega^-(U) = \emptyset$, and so $c(\omega^-(U)) = 0$, and Gale's condition can be written:

$$0 \geq |U \cap X| - |U \cap Y|$$

and this condition is really fulfilled since we have:

$$|U \cap Y| \geq \big|N(U \cap X)\big| \geq |U \cap X|$$

the second inequality being given by Hall's condition.

8.5.3 König's theorem

Let us recall this other classic matching theorem (Chapter 7, theorem 7.3): *if a graph is bipartite, then* $\nu(G) = \tau(G)$, where $\nu(G)$ is the maximum of edges of a matching and $\tau(G)$ is the minimum of vertices of a transversal of G.

This result is also an equality expression between the value of a maximum flow and the capacity of a minimum cut in a network. Let us build the following network R from bipartite graph $G = (X, Y, E)$. Each edge of G is directed from X to Y and receives capacity ∞. On the one hand we add a vertex s, which will be the single source of the network, and an arc (s, x) with capacity 1 for each $x \in X$. On the other hand, we add a vertex t, which will be the single sink of the network, and an arc (y, t) with capacity 1 for each $y \in Y$. It is not very difficult to verify what follows:

- A flow of R defines a matching of G and the value of the flow is equal to the number of edges of the matching (consider the arcs of flow equal to 1).

- A finite cut of R defines a transversal of G and the capacity of the cut is equal to the number of vertices of the transversal. Specifically, if the cut of R is defined by the set Z, the transversal of G is $(X \setminus Z) \cup (Y \cap Z)$.

Thus, the value of a maximum flow of R is equal to the number of matching $\nu(G)$ and the capacity of a minimum cut of R is equal to the minimum transversal $\tau(G)$. Ford-Fulkerson's theorem then directly gives the equality $\nu(G) = \tau(G)$ of König's theorem.

The theory of flows in transportation networks presents a remarkable algebraic extension with the concept of *tension*, dual to that of flow, defined by a *potential function*, as in electricity.

8.6 Exercises

+8.1. Show that in a network with a source s and a sink t, and for which all arcs have a capacity > 0, there is no other flow than the zero flow if and only if there is no path from s to t in the network digraph.

8.2. Find a maximum flow in the network in Figure 8.1. (Deal first with the network in Figure 8.2.)

8.3. Re-apply the maximum flow algorithm to the network in Figure 8.4, using the vertex rule: "first labeled, first explored". Compare the number of iterations.

+8.4. Let R be a given network defined on a digraph G with a flow f. We consider the *unsaturating digraph*, weighted digraph H defined in the following way: it has the same set of vertices as R and there is in H an arc b from vertex x to vertex y if and only if there is in G an arc a directed either from x to y with $f(a) < c(a)$, or from y to x with $f(a) > 0$. In the first case, arc b is weighted by $c(a) - f(a)$ and in the second case by $f(a)$. Using this digraph, reformulate the algorithmic search for an unsaturated path. Interpret the application of the "first labeled, first explored" rule and its consequences for the unsaturated paths obtained.

8.5. In the network in Figure 8.1, justify the answer for the existence of a feasible flow with the stocks and demands given in the table below (for some cases use the result of exercise 8.2):

$\sigma(s_1)$	$\sigma(s_2)$	$\sigma(s_3)$	$\tau(t_1)$	$\tau(t_2)$	Answer
3	5	2	8	3	no
3	8	2	8	5	no
5	5	2	8	4	no
5	5	2	10	2	no
3	9	0	8	4	yes

Chapter 9

Euler Tours

Organizing a "tour", going through each street of a city to distribute the mail or to collect garbage, is a classic problem. The easiest way *a priori* would be to try to go through each street only once, but this is not always possible, as we will see.

The graphs under consideration in this chapter are undirected and are not supposed to be simple (loops and multiple edges are allowed). It is possible to extend what follows to digraphs.

9.1 Euler trails and tours

A *Euler trail* of a graph G is a trail containing all the edges of G, that is which goes exactly once through each edge of G. A *Euler tour* of G is a closed Euler trail. A graph is called a *Eulerian* graph if it has a Euler tour. The origin of this concept dates back to the 18th century with the famous "Königsberg bridge problem", well before graph theory existed as such. In this town, today called Kaliningrad, a Russian enclave in the Baltic countries, there are seven bridges built on the river, as shown in Figure 9.1. At that time a citizen wondered if it was possible to cross all of those bridges once each in the course of a single walk bringing them back to their starting point. The mathematician Euler, interested in this problem, made explicit the condition proving it was not possible (the word "Eulerian" used today comes from his name of course). Modeling this problem with a graph is easy (see Figure 9.1). A vertex is associated with each neighborhood of the town and an edge is associated with each bridge, joining the two vertices associated with the neighborhood linked by this bridge. Touring the bridges, as a Königsberg citizen wanted to do, amounts to the existence of a Euler tour in the associated graph.

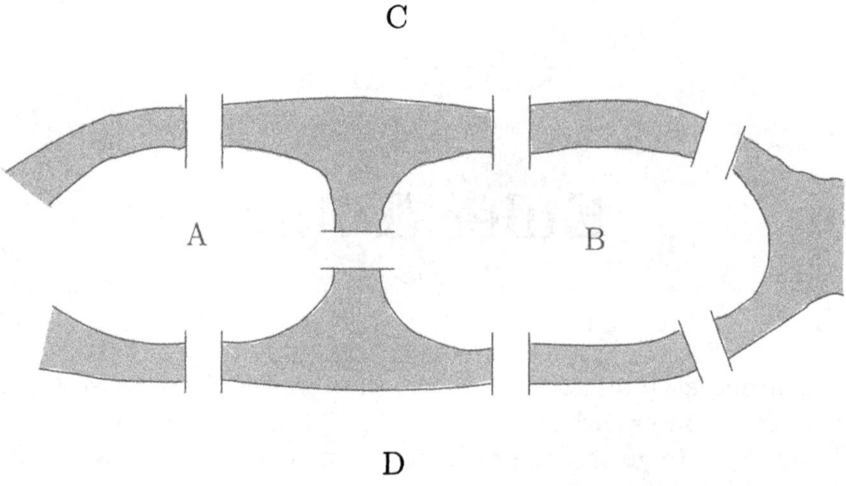

Königsberg bridges

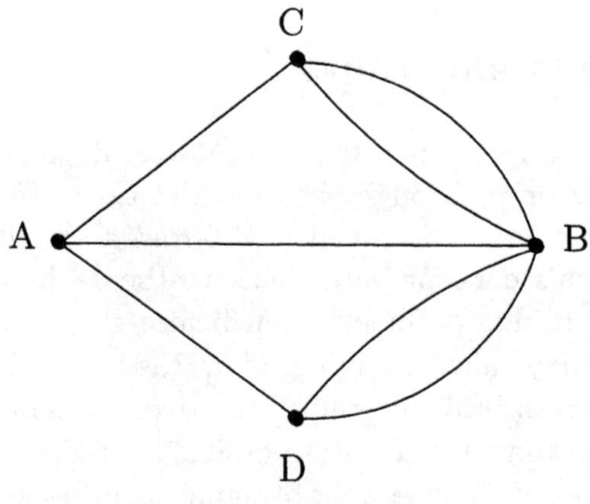

associated graph

Figure 9.1. *Königsberg bridge problem and its graph model*

A different version of this problem is known to the many pupils who have been asked if it is possible to draw various small shapes with a single line (with no repetition of the line). Two examples are given in Figure 9.2 (to be verified). Modeling using a graph is obvious and this time the problem † does not need a return to the starting point. This answer is linked to the existence of a Euler trail in the graph.

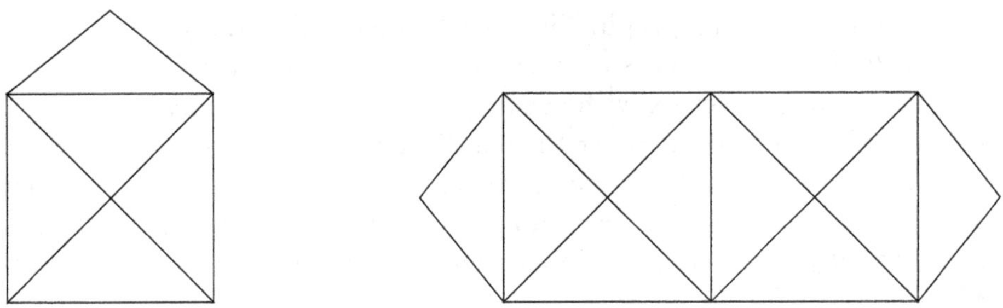

Figure 9.2. *Two forms that can be drawn using a single line*

9.1.1 Principal result

THEOREM 9.1 (Euler). *A connected graph G is Eulerian if and only if every vertex has an even degree.*

Proof. Let m be the number of edges of G. The case $m = 0$ being trivial, we can suppose $m \geq 1$. The necessary condition is almost immediate. Indeed, in following a Euler tour, which we suppose to exist, we necessarily visit each vertex of the graph an even number of times since each time we reach a vertex, we then leave it by a different edge, by definition. The sufficient condition is less simple but the proof that we will give will prefigure a recursive search algorithm of a Euler tour in a graph. We can reason by induction on m. The property is true for $m = 1$: because of the hypothesis on the degrees, the graph is reduced to one vertex with a loop which defines a Euler tour. Let us suppose $m > 1$ and the property true for any graph with a number of edges $< m$. The minimum degree δ of G is ≥ 2, because the graph is connected, not reduced to an isolated vertex, and each vertex has a non-zero even degree. We deduce the existence of a closed trail in a constructive manner: we start from any vertex and go each time to a neighbor of the last vertex visited without returning to the one just under consideration previously. This is always possible since $\delta \geq 2$. We continue until we find a vertex which has been previously visited, which closes a trail C of G (in fact, a cycle). If this trail is not a Euler tour, we consider the connected components which are not reduced to an isolated vertex of graph $G - E(C)$ ($E(C)$ designating the set of the edges of C). Because of the hypothesis that C is not a Euler tour, there is at least one such component. Let $H_1, \ldots, H_k$ be these components. By the induction hypothesis, each H_i, $i = 1, \ldots, k$ is a Eulerian graph. Thus, let C_i be a Euler tour of H_i, for each $i = 1, \ldots, k$. Since graph G is connected, trail C meets each C_i, that is has at least one common vertex with each C_i. Let x_i be one of these vertices. It is then possible to define a Euler tour of G by inserting C_i into C at vertex x_i for each $i = 1, \ldots, k$. To be more specific, and in a constructive manner, this trail may be defined by searching C and, when meeting x_i, leaving C in order to fully search C_i, returning then to x_i to continue searching trail C where it was left. □

COROLLARY 9.1. *A connected graph has a Euler trail if and only if the number of its vertices of odd degree is* ≤ 2.

Proof. Remember that any graph has an even number of vertices of odd degree. If the graph considered has no odd degree vertices, it is Eulerian and has a Euler tour which is also a (closed) Euler trail. If graph G has two vertices of odd degree, let them be x and y; consider graph G' obtained by adding to G an edge joining x to y. Graph G' verifies the hypotheses of the theorem, therefore it allows a Euler tour. This tour reduced to G, that is minus the edge added to G, yields a Euler trail with endvertices x and y. Finally, if the considered graph has more than two vertices of odd degree, it cannot have a Euler trail: indeed such a trail, if it is not closed, has for ends vertices which are the only vertices of odd degree, therefore there cannot be more than two in the graph. □

NOTES. 1) In a graph with two vertices of odd degree, any Euler trail necessarily has these two vertices for ends. Thus, the examples in Figure 9.2 have to be drawn with a line starting at one of the two vertex "nodes" of odd degree and ending at the other.

2) The connectedness hypothesis in theorem 9.1 and its corollary are not necessary. Indeed, a graph may have isolated vertices while being Eulerian. The correct hypothesis instead of connectedness is that the graph has at

most one connected component which is not an isolated vertex (but generally we are not interested in components reduced to one vertex).

9.2 Algorithms

The proof of theorem 9.1 directly inspires the following recursive function, which, given a graph G, returns a Euler tour of that graph. The graph considered must be connected and Eulerian. Note that it is simple to first verify that the graph is truly Eulerian by checking that the degrees of its vertices are all even, in accordance with theorem 9.1. The algorithm is expressed in the form of a recursive function. The graph is given by lists of incident edges at each vertex, with for each edge the datum of its other endvertex. A connected component of $G - E(C)$ is here called "non-trivial" if it is not reduced to an isolated vertex. These are the only components of $G - E(C)$ worth considering in the recursive calls (following the proof of theorem 9.1).

```
function Euler1(G) return closed trail;
begin
  construct_closed_trail(C,G);
  if E(G)\E(C) = ∅ then return C; end if;
  determinate_components(H1,...,Hk);
  for i in 1..k loop
    Ci:= Euler1(Hi);
  end loop;
  insert(C,C1,...,Ck);
  return C;
end Euler1;
```

Some of the statements are made explicit as follows:

```
construct_closed_trail(C,G):
```

Choose (any) vertex x_0 of G, if $d_G(x) \neq 0$ then consider an edge incident to x_0, take its other endvertex x_1 ($x_0 = x_1$ is possible, there can be a loop in x_0), then start again with x_1, considering an edge incident to x_1 different from the previous one. Continue in this manner, always considering an edge which has not been previously considered. Stop when a vertex previously encountered is visited again. A closed trail is then defined by the sequence of the edges considered. Note that it is also possible to continue as long as there is an edge not yet considered incident to the current vertex. The process will then necessarily stop at the initial vertex x_0 (with the hypothesis that the vertices of the graph are all of even degree).

`determinate_components(H₁,...,Hₖ):`

This determines the non-trivial connected components $H_1, \ldots, H_k$ of $G - E(C)$: the connected components of $G - E(C)$ can be determined by a standard algorithm of connectedness (for example by a search of the graph). We must then eliminate those reduced to an isolated vertex. This statement is only executed in the case $E(G) \setminus E(C) \neq \emptyset$, a hypothesis which insures the existence of at least one component which is not reduced to an isolated vertex. The case $E(G) \setminus E(C) = \emptyset$ defines the exit condition of the recursive calls.

`insert(C,C₁,...,Cₖ):`

This operation for inserting trails C_i in C is explained in the proof of theorem 9.1. It assumes knowledge of a common vertex of C and of the cycle to be inserted. In a concrete manner, for each i, such a vertex x_i common to C_i and C is found while searching C. It is enough then to locate on this vertex the sequence defining trail C_i, and then to continue searching trail C after x_i. At the level of programming, it is best to represent C and C_i by *linked lists.* This will make the insertion operation, which amounts simply to assignments of pointers, easy.

9.2.1 Example

See Figure 9.3. At the top is the Eulerian graph considered, with vertices numbered from 1 to 12 and edges labeled from a to x. In the following the trails considered are simply described by the sequences of their edges. The closed trail considered first, $C = (a, b, c, d, e, f, g, h, i)$, is shown in bold. At the bottom is the graph $G - E(C)$. It has two non-trivial connected components, H_1 and H_2, respectively induced by the sets of vertices $\{2, 4, 6, 8, 10, 11, 12\}$ and $\{3, 5, 7, 9\}$. Euler tours for these components are, for H_1:

$$C_1 = (o, u, v, r, p, q, x, t, s, w, n)$$

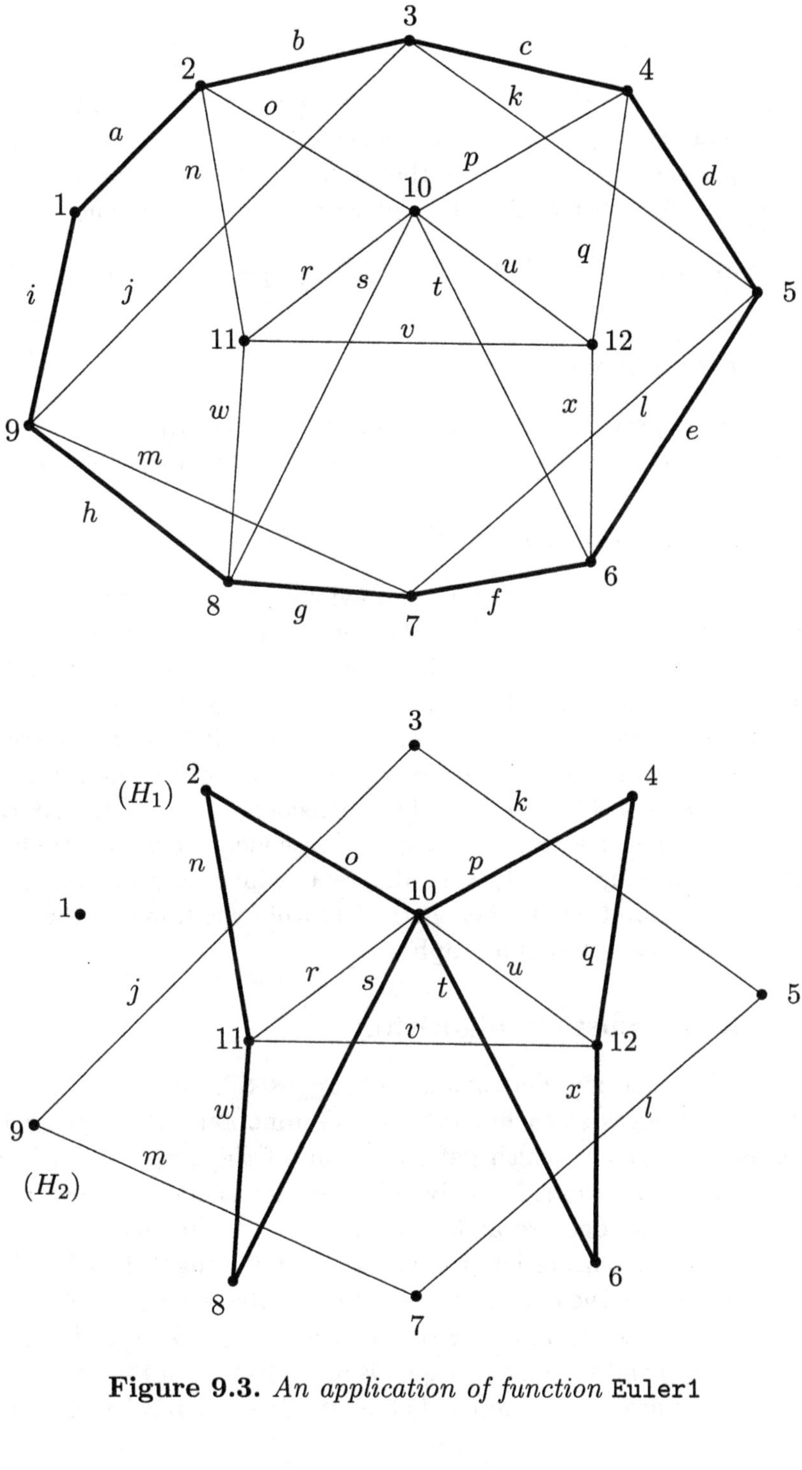

Figure 9.3. *An application of function* `Euler1`

and for H_2:

$$C_2 = (k, l, m, j)$$

Trail C_2 is obtained directly (no new recursive call). Closed trail C_1 is obtained recursively by inserting, at vertex 10, closed trail (u, v, r) in closed trail (o, p, q, x, t, s, w, n) (in bold in the subfigure at the bottom). Finally, by inserting C_1 in C at vertex 2 and C_2 at vertex 3, we obtain the Euler tour:

$$(a, o, u, v, r, p, q, x, t, s, w, n, b, k, l, m, j, c, d, e, f, g, h, i)$$

9.2.2 Complexity

With appropriate implementation, this algorithm has a linear complexity, $O(m)$, where m is the number of edges of the graph considered.

9.2.3 Elimination of recursion

Some observations, made from the previous recursive algorithm, lead, by elimination of recursion, to an iterative algorithm. The crucial observation is that it is possible to insert in C the intermediary closed trails found along the way as they are built. This saves having to do this work during the returns of the recursive calls. The initial closed trail C is thus augmented as work progresses and ends (before any recursive return) as a Euler tour when all the edges of the graph have been considered. This of course requires more technical specifications but it gives the general idea and leads to an algorithm which is apparently very different, which we will now give and which has been found by another method (within the frame of the study of different types of searches of a graph).

9.2.4 The Rosenstiehl algorithm

The graph under consideration is still supposed Eulerian and is given by lists of incident edges at each vertex. In the function `Euler2` given below, `C` represents a sequence which defines a trail of the graph and which will be at the end a Euler tour. Initially this sequence is empty. When a vertex or a vertex and an edge are added to `C`, it is just after what has already been placed, and in that order (the vertex then the edge). In what follows, variable `z` represents the current vertex and variable `e` an edge. A stack `S` is used which can stock ordered pairs of the form `(e,y)`, where `e` represents an edge of the graph and `y` a vertex, which is an endvertex of the edge. Initially the stack is assumed to be empty. Let us recall a technical point: the `pop` procedure only amounts to removing the element which is at that moment at the top of the stack (without returning it). An edge is said to be *free* if it has not yet been considered. Initially any edge is free. A vertex is said to be *open* if it has at least one free incident edge. The addition at the end of the running of vertex `r` in `C` closes this trail (which having started at vertex r can only close at that same vertex).

The expression of this algorithm may seem complicated, but its running is remarkably simple: we push as long as there is a free incident edge at the current vertex, otherwise we pop until it is the case again or until the stack is empty. *The edges leave the stack in the order of a Euler tour of the graph.* In addition, this algorithm clearly has a linear complexity, $O(m)$.

```
function Euler2(G) return closed trail;
begin
  r:= a vertex of G;
  z:= r;
  loop
    while z is open loop
      e:= a free edge incident to z;
      mark e not free;
      y:= the other endvertex of e;
      push(S,(e,y));
      z:= y;
    end loop;
    add z,e in C;
    pop(S);
    exit when is_empty(S);
    (e,z):= stack_top(S);
  end loop;
  add r in C;
  return C;
end Euler2;
```

Let us go over the example in Figure 9.3 once more, in Figure 9.4. Only the edges are indicated in the stack (the corresponding endvertices are directly visible in the figure). The removals of the stack correspond, in that order, to the edges of the Euler tour obtained, different from the previous one and which is:

$$(i, m, l, k, j, h, s, t, x, u, r, v, q, p, o, n, w, g, f, e, d, c, b, a)$$

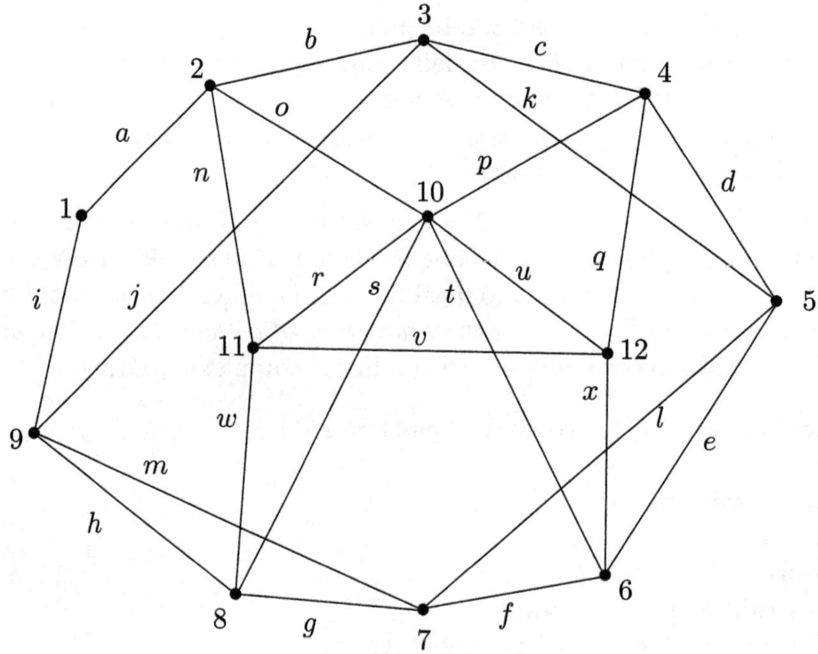

stack:	removals of the stack:
$abcdefghi$	i
$abcdefgh$	
$abcdefghjklm$	m, l, k, j, h
$abcdefg$	
$abcdefgwnopqxts$	s, t, x
$abcdefgwnopq$	
$abcdefgwnopqvru$	$u, r, v, q, p, o, n, w, g, f, e, d, c, b, a$
$\emptyset$	

Figure 9.4. *An application of function* `Euler2`*. The stack is represented with only the edges concerned (the corresponding endvertices are seen directly on the graph). The vertex initially chosen is 1*

9.3 The Chinese postman problem

The so-called *Chinese postman problem* (thus called because of the nationality of the first person to be interested by it) is typically an optimal tour problem. It means, for example, touring all streets of a town while minimizing the total distance covered.

In graph terms, the general problem is as follows: given a graph G with edges weighted with values ≥ 0, we have to find a *closed walk* of G going at least once through each edge of G, and for which the sum of the values of its edges is as small a possible. Let us specify that in general this closed walk is not a closed trail, which means that an edge of the graph may appear in it several times. Thus, the minimum total value can be greater than the sum of the values of all the edges of the graph because of the edges covered several times. In the particular case of a Euler tour, we have equality, since a Euler tour is then a solution to the problem. When the graph is not Eulerian, while still supposed connected, a closed walk solution must necessarily go more than once through at least one edge of the graph. It is all a matter of choosing these multiple edges in such a way as to minimize the sum of their values. Let us make this more specific. In what follows we will write CPP for "Chinese Postman Problem".

Let $G = (X, E)$ be a connected graph weighted by $v : E \to \mathbb{R}^+$. Let us put Y the set of the odd (degree) vertices of G, supposed non-empty. We first have:

– *A closed walk of G corresponds to a set F of edges defined on X, which is disjoint from E and which verifies the two following properties:*

1) *any edge of F is parallel to an edge of E,*

2) *graph $H = (X, F)$ has the same set, Y, of odd vertices as G.*

Given a closed walk D of G, consider the set F composed in the following manner: for any edge e of G appearing $m(e) > 1$ times in D, put in F $m(e) - 1$ edges parallel to e. It is easy to verify that this set F verifies the two preceding properties. In particular the second one is linked to the fact that graph $G + F = (X, E \cup F)$ is Eulerian: it is indeed possible to define a Euler tour of $G + F$ by searching walk D and by doubling each edge of G met a second time, which comes back to construct $G + F$. The vertices of $G + F$ are therefore of even degree, which implies that graphs G and H have the same odd-degree vertices (an odd vertex in one set is also odd in the other because the sum of these degrees is even). Conversely, given a set F verifying the preceding properties, graph $G+F$ is Eulerian because it is connected and its vertices are even-degree vertices, and a Euler tour of this graph defines, in an obvious manner, a closed walk of G (by repeating in G the edges doubled in $G+F$).

Let us now suppose that the edges of set F are weighted, each with the same value as the edge of G to which it is parallel. We add the third following property:

3) the value $v(F) = \sum_{e \in F} v(e)$ *is minimal.*

This property is equivalent to the next one, which defines an optimal solution of the CPP in G, where D is the closed walk associated with set F as previously:

3') the value $v(D) = \sum_{e \in D} v(e)$ *is minimal.*

Let us specify that the sum which defines $v(D)$ is taken on the sequence of the edges of walk D, that is, an edge is counted as many times as it appears in this walk. This equivalence between properties 3 and 3' comes directly from the equality $v(D) = v(E) + v(F)$ and from the fact that the value $v(E) = \sum_{e \in E} v(e)$ is constant.

We can reformulate what has been described above by saying that a solution D of the CPP in G corresponds to a set F of edges which verifies the preceding properties 1, 2 and 3. We are going to characterize set F in a manner which will show how to obtain it. Let us make d the distance function in (weighted) graph G, that is, $d(x, y)$ is the distance between the vertices x and y. Let K_Y be the complete graph defined on the vertices set Y (odd vertices of G), weighted on the edges by d, which means that edge xy of K_Y has the value $d(x, y)$. We have:

– *Set F verifies the preceding properties 1, 2, and 3 if and only if it corresponds to a minimum matching of K_Y.*

Let us further explain this correspondence between F and a matching of K_Y. Let us suppose that F verifies properties 1, 2, and 3. Let us consider in $H = (X, F)$ a vertex x_1 of odd degree. According to an easy result (proposed as an exercise in Chapter 1, exercise 1.3), there is another vertex of odd degree in H; let it be x'_1, linked to x_1 by a walk μ_1, which we can suppose to be a path. Let us put $H_1 = H - F(\mu_1)$, where $F(\mu_1)$ designates the set of the edges of μ_1. In H_1, vertices x_1 and x'_1 are even-degree vertices. If H_1 still has an odd-degree vertex, then we start again with this graph, by defining a path μ_2 which links two odd-degree vertices x_2 and x'_2. We continue until at last graph H_k is reduced to two odd vertices x_k and x'_k linked by a path μ_k. We thus have k paths of H (H_i are spanning subgraphs of H) which are pairwise disjoint for the edges, and which "match" the odd-degree vertices of H; these are also those of G since they are the same vertices according to property 2 (remember that a graph always has an even number of odd vertices). It easily follows from the minimality of $v(F)$ that each path μ_i must be a shortest path between the vertices x_i and x'_i which it is linking (otherwise it would suffice to replace this path by a shorter one). It is therefore possible to associate with each path μ_i the edge $x_i x'_i$ of K_Y, with its value $d(x_i, x'_i)$. This thus defines a matching M of K_Y associated with the set of the preceding paths. We have $l(\mu_i)$, the length of path μ_i:

$$v(F) \geq \sum_{i=1,\dots,k} l(\mu_i) = \sum_{i=1,\dots,k} d(x_i, x'_i) = v(M)$$

Conversely, given a matching M of K_Y, it is easy to associate with it a set F verifying the above properties 1 and 2 by considering some parallel edges for all the edges of the paths which correspond in G to the edges of M. With this set F associated with a matching M of K_Y, we have the equality $v(F) = v(M)$. Thus the value of $v(F)$ is minimal, that is it verifies property 3, if and only if set F is associated with a minimum matching M.

We thus see how to obtain a set F of edges corresponding to a solution of the CPP in G from a minimum matching of K_Y. All this leads us to the following algorithm.

9.3.1 The Edmonds-Johnson algorithm

Let us describe this algorithm step by step.

Step 1: Find for each pair of vertices of odd degree of G a shortest path (in the sense of the weighting of G) which links these vertices.

Step 2: Build a complete weighted graph K of which the vertices are the odd vertices of G, and of which each edge is weighted by the length of the shortest path linking its ends found in step 1. Determine a minimum matching M of K.

Step 3: Build a weighted graph $\hat{G}$ by adding in G for each edge of M the edges of the shortest corresponding path (found in step 1). Each added edge is added independently of those already existing. The parallel edges thus created are given the same value by v as those of G.

Step 4: Find a Euler tour of $\hat{G}$.

Graph $\hat{G}$ corresponds to graph $G + F$ seen above. The Euler tour of $\hat{G}$ defines, as we have already seen, a solution of the CPP in G. Let us comment on these different steps. Step 1 can be accomplished by repeatedly using Dijkstra's algorithm (Chapter 6). To find a matching of minimal value in K during step 2, there is a general algorithm not dealt with in this book and more general than the one given in Chapter 7 which is limited to bipartite graphs. Step 3 is direct. Finally, step 4 is solved by the algorithm previously given in this chapter.

9.3.2 Complexity

From a complexity point of view, it is possible to verify that this algorithm is, as a whole, polynomial since that is the case for each of its steps. Thus, the Chinese postman problem is solved by a "good" algorithm.

9.3.3 Example

The number of examples we can give is limited here by the fact that we have not given any general algorithm for finding a matching of minimal value in K during step 2. Nevertheless it is possible to deal with the case where there are few vertices of odd degree. In this case, it is possible to find such a matching directly by considering all possible matchings, which are not numerous (for example, 1 for 2 vertices, 3 for 4 vertices).

In the example given in Figure 9.5, there are only two vertices of odd degree. Its treatment is therefore trivial (see in exercise 9.4 another case with four vertices of odd degree). At the top of Figure 9.5, see graph G with, in bold, a shortest path joining the two vertices of odd degree x_0 and x_2. This path was obtained using Dijkstra's algorithm applied from vertex x_0. At the bottom, observe graph $\hat{G}$ obtained by doubling in G the edges of the previous path. All the vertices of this graph are even, therefore it is Eulerian. It remains to find a Euler tour of $\hat{G}$ with one of the algorithms given above. This tour defines a solution to the Chinese postman problem in G (a solution tour goes twice through each of the edges of G in bold).

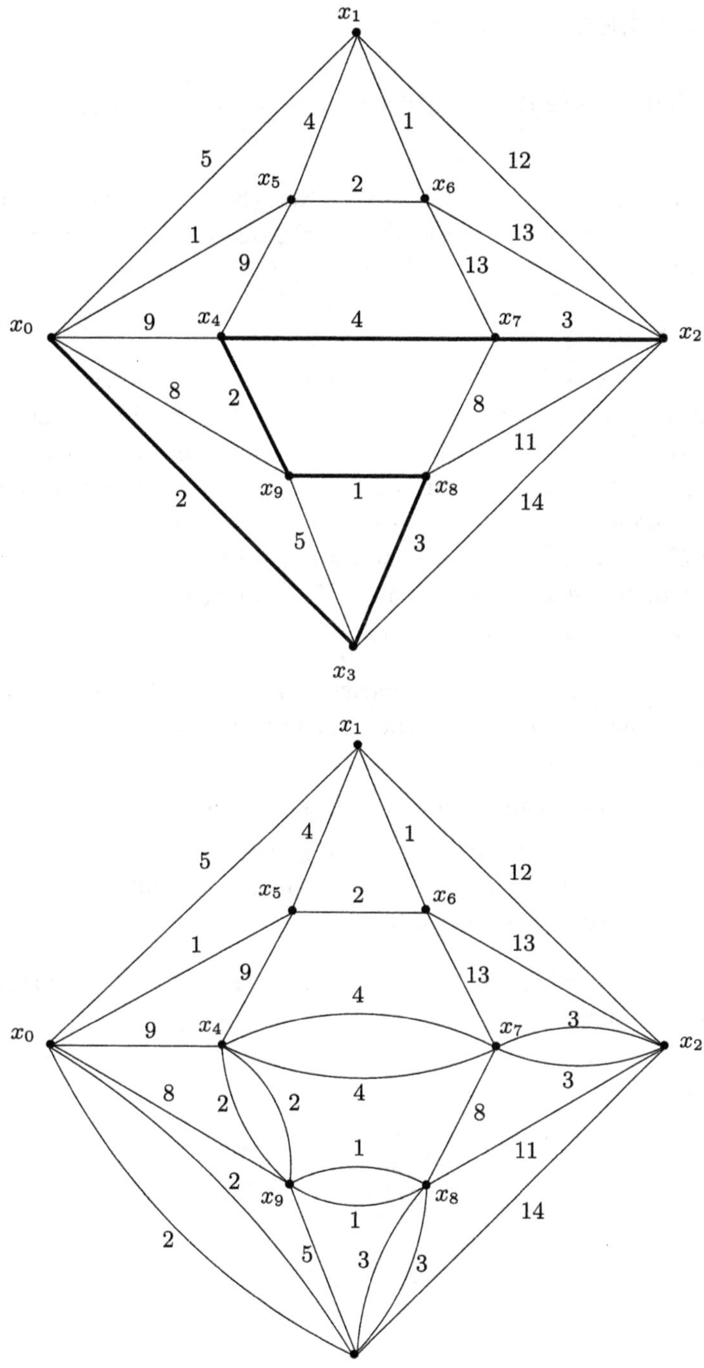

Figure 9.5. *An example of a solution to the Chinese postman problem*

9.4 Exercises

9.1. Show that if a graph G is Eulerian then each of its blocks is Eulerian. Is the converse true?

9.2. Show that if a graph G has no vertices of odd degree, then the set of its edges can be partitioned into classes, each being the set of the edges of a cycle.

*9.3. *(Fleury's algorithm)*

Given a connected graph G, let us consider the following construction of a trail. We randomly choose a first vertex x_0. Let us suppose the walk $(x_0, e_1, x_1, \ldots, e_k, x_k)$, with $k \geq 0$, is already built. Let $e_{k+1} = x_k x_{k+1}$ (if it exists) be an incident edge to x_k not yet taken, and which, *except if it is impossible to do otherwise*, is not a cut edge of graph $G_k = G - \{e_1, \ldots, e_k\}$ (spanning subgraph of G induced by the edges which are not yet taken). We start again with the new walk $(x_0, e_1, x_1, \ldots, e_k, x_k, e_{k+1}, x_{k+1})$.

a) Show that, with the hypothesis that G is Eulerian, it is always possible to build this trail and that it always ends back at the starting vertex x_0.

b) Show that a Euler tour of G has thus been constructed.

c) Reflect on how to give an algorithmic representation of this construction (in particular recognition of cut edges) and on its complexity (polynomial?).

9.4. Apply the algorithm solving the Chinese postman problem to the graph in Figure 9.6.

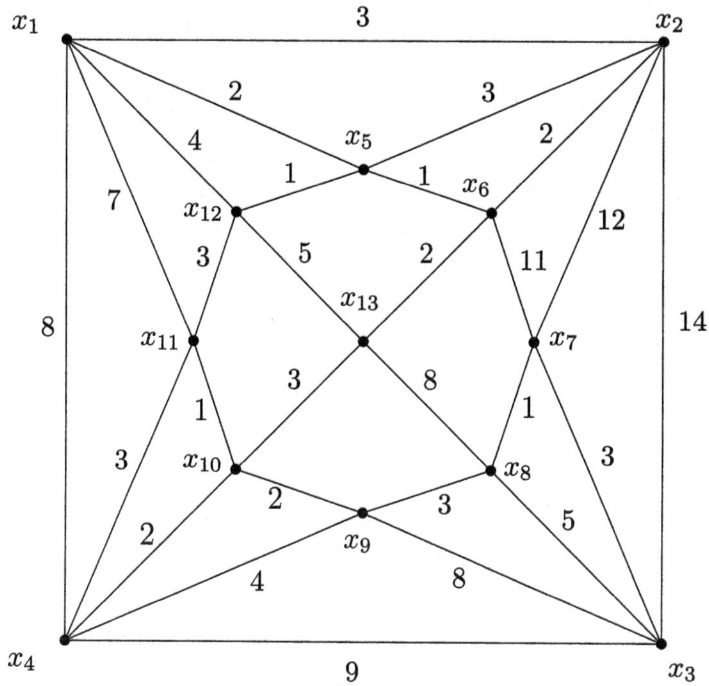

Figure 9.6. *Exercise 9.4*

Chapter 10

Hamilton Cycles

After the Euler tours studied in Chapter 9, the problem now under consideration is, for example, to go once and only once through all the cities of a given set, returning to the starting point at the end of the tour. In graph terms this means looking for Hamilton cycles. This represents a theoretical and algorithmic difficulty significantly greater than the Euler tours studied in the preceding chapter. In a weighted graph, we also take into account the values of the edges followed going from one vertex to another. By trying to minimize the sum of these values, we will define what is probably the most famous problem in combinatorial optimization, the "traveling salesman problem", which we will study in this chapter.

10.1 Hamilton cycles

The graphs considered in this chapter are undirected and are supposed simple. A *Hamilton cycle* of a graph G is a cycle going through all the vertices of the graph. Because it is a cycle, it will only go through each vertex once (counting only once the first and last vertex, which is the same vertex since it is a cycle). A graph G is a *Hamiltonian graph* if it has a Hamilton cycle.

This concept and its name historically come from a "world tour" problem considered by the mathematician Hamilton in the 19th century. This problem consisted of finding a Hamilton cycle in the graph in Figure 10.1.

We also define the concept of a *Hamilton path*: a path going through all the vertices of a graph. From a "tour" point of view there is no longer the constraint of having to return at the end to the starting point. A classic

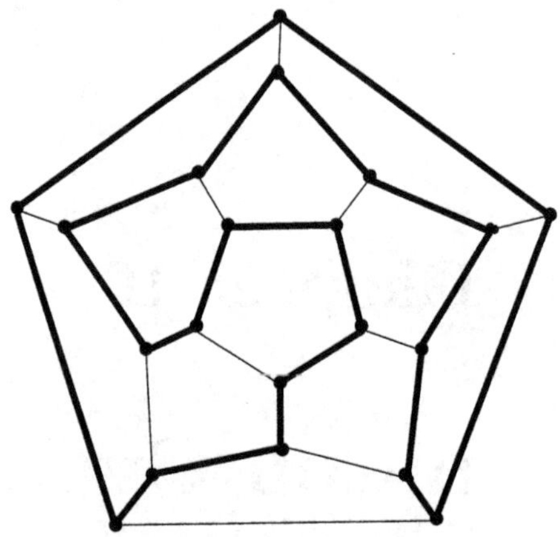

Figure 10.1. *A Hamilton cycle (bold lines) in the "world tour" graph considered by Hamilton, in the form of a regular dodecahedron (the vertices represent large cities of the world)*

example of this problem is the "knight's move" on the chessboard: can the knight cross every possible chessboard square only once (using the habitual movements for a knight)? There are multiple solutions for this problem.

10.1.1 A few simple properties

The following properties are easy to verify:

1) A complete graph K_n is Hamiltonian for any $n \geq 3$. This is in general † the case with many different cycles, meaning $\frac{(n-1)!}{2}$ (remember that two cycles are not considered different if they only differ by their cyclic sequences of vertices which define them).

2) The problem of the existence of a Hamilton path in a graph G can come down to the existence of a Hamilton cycle in G', where G' is the graph obtained by adding to G a vertex u joined by an edge to each of the vertices of G.

3) A bipartite graph $G = (X, Y, E)$ can only be Hamiltonian if $|X| = |Y|$. In particular if $|X \cup Y|$ is odd, G is not Hamiltonian.

4) A Hamiltonian graph is necessarily 2-connected as is easy to see. However, that is not enough as is shown by the graph in Figure 10.2,

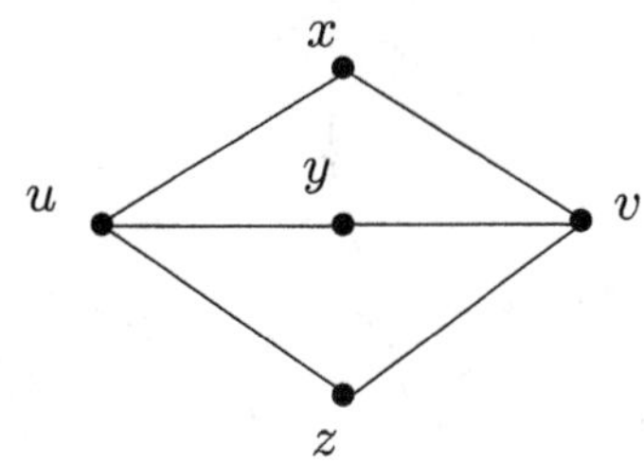

Figure 10.2. *A non-Hamiltonian 2-connected graph*

a complete bipartite graph $K_{2,3}$, which is not Hamiltonian in keeping with the preceding property.

The following proposition gives a necessary condition for existence of a Hamilton cycle which may be useful.

PROPOSITION 10.1. *If graph $G = (X, E)$ is Hamiltonian then for any $S \subseteq X$ such that $S \neq \emptyset$ and $S \neq X$, we have, noting $p(G - S)$ the number of connected components of subgraph $G - S$:*

$$p(G - S) \leq |S|$$

Proof. A Hamilton cycle of G shares the set of the vertices of $G - S$ in at most $|S|$ non-empty subsets corresponding respectively to the pieces of the cycle cut by the vertices of S. Each of these subsets is in a connected component of $G - S$ and each connected component contains one or more of these subsets. The number of these subsets therefore must be at least equal to the number of connected components. □

This necessary condition for existence of a Hamilton cycle is not sufficient as is shown in the case of a Petersen graph (Figure 10.3) which is not Hamiltonian. In fact, there is no simple necessary and sufficient condition for a graph to be Hamiltonian. However, we have many sufficient conditions, in particular based on degrees conditions. The following theorem is a typical example.

THEOREM 10.1 (Ore). *A simple graph $G = (X, E)$, with $n \geq 3$ vertices, is Hamiltonian if $d(x) + d(y) \geq n$ for any two non-neighboring vertices $x, y \in X$.*

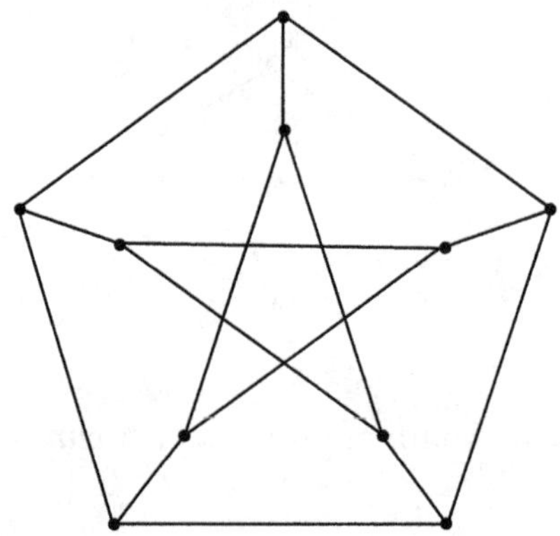

Figure 10.3. *Petersen graph*

The problem of the existence of a Hamilton cycle is therefore not easy from a theoretical point of view. Nor is it simple from a practical algorithmic one, since it is **NP**-complete. This difficulty will be found again in the more general frame of an optimization problem.

10.2 The traveling salesman problem

The *traveling salesman problem* encompasses the problem of the existence of a Hamilton cycle in a graph. In a practical form, the problem is that of a traveling salesman who has to tour a certain number of cities, leaving from one of them, going exactly once through the others, to return at the end to the city from which he departed. This traveler wants to minimize the total distance covered.

In graph terms, the problem is set in its most general form in the following manner. Let K_n be a complete graph, $G = (X, E)$, weighted by the mapping $v : E \rightarrow \mathbb{R}^+$ (real numbers ≥ 0). The question is to find a Hamilton cycle of G of which the total value, defined as the sum of the values by v of its edges and denoted by $v(C)$, is as small as possible. In the case of "the traveling salesman", the weighting v corresponds to the distances between the cities. Our frame here, in terms of graphs, is more general and v is random; in particular it does not verify the *triangular inequality* specific to distances. However, we will still consider this interesting particular case later.

10.2.1 Complexity of the problem

As for any optimization problem, the goal here is to find an element which is an extremum following a certain measure, in a set which has in general so many elements that it is impossible in practice to consider them all. A search which would consider all cases is impracticable. For example, in a complete graph with n vertices there are $\frac{(n-1)!}{2}$ distinct Hamilton cycles. It is easy to understand that as soon as integer n is large, considering all the cycles in order to find one that is less than or equal to all the others is not possible even using the most powerful machine. To define the difficulty of the traveling salesman problem, let us show how it contains the problem of the existence of a Hamilton cycle in a graph, a problem which we already mentioned as **NP**-complete.

Let $H = (X, F)$ be a graph with n vertices. Let us associate with this graph the complete graph $G = (X, E)$ with the same set of vertices and weighted by putting: $v(xy) = 1$ if $xy \in F$ and $v(xy) = 2$ if $xy \in E \setminus F$. It is easy to verify that a Hamilton cycle of H corresponds in G to a Hamilton cycle of total value n, and conversely. So the existence of a Hamilton cycle in H corresponds to the existence of a Hamilton cycle in G with a total value $\leq n$ (in fact equal to n). Thus, an algorithm which solves the traveling salesman problem in G also solves at one blow the problem of the Hamilton cycle in H. In what we just did, which is called the *reduction* of a problem to another (see Appendix B), we have used the *decision* problem associated with the traveling salesman problem, which can be put as follows (and which is, in fact, of an equivalent algorithmic difficulty): given a complete weighted graph G and an integer k, is there in G a Hamilton cycle with a total value $\leq k$? This problem, which contains the problem of the existence of a Hamilton cycle, is therefore **NP**-complete.

10.2.2 Applications

We can wonder what the practical interest of the traveling salesman problem is. Indeed, there are probably very few traveling salesmen who organize their tours in such simple terms. They usually have a lot more constraints to take into account such as, for example, the need to be in a certain city on a certain date to meet someone. The constraint of the shortest distance covered is therefore not always the most pertinent in practice. However, we can cite real applications of the traveling salesman problem. The first one, which is quite obvious, concerns the programming of the articulate arm of a robot which needs to cover certain points of a sheet for welding. For speed, and possibly in order to economize and limit the wear out of machine tools, we might want to minimize the total distance covered by the extremity of the arm of this robot. Modeling this problem into a "traveling salesman problem" is direct.

The other application considered here is less direct. In a paint workshop, a number of blends have to be prepared each day. A mixer prepares the blends but going from one mixing to another requires adjusting the mixer depending on the blends to be made. We know the time necessary to adjust the machine to go from one mixing of a pair of colors to another. This time is assumed to be symmetric with regard to the order of the two blends. The problem is to define the order of passage of the blends in the mixer in order to minimize the total time spent in adjusting it. If we want to finish with the machine set for the first mixing, and if each mixing is only done once, the problem can once again be modeled as a traveling salesman problem by considering the graph of which the vertices are the blends to be prepared, each edge being weighted by the time necessary to go from one mixing to another. A Hamilton cycle with a minimal total value answers the question.

10.3 Approximation of a difficult problem

We will from now on write TSP for "traveling salesman problem". The TSP is therefore **NP**-complete. What can be done with an **NP**-complete problem? The hope of finding a good algorithm, that is a polynomial one, is low since it would be proof that $\mathbf{P} = \mathbf{NP}$. On the other hand, if $\mathbf{P} \neq \mathbf{NP}$ such an algorithm does not exist at all.

When motivated by the needs of an application, a natural course is to try to obtain at least an approximate solution, that is one that may not be optimal in general but which is not too far from an optimal solution. This objective may already be ambitious. For example, a "greedy" strategy, which works well for the problem of the minimum spanning tree (Chapter 2), fails completely here. This strategy for the TSP consists of choosing to go each time to the nearest city not yet visited. We can give infinite families of cases for which the result obtained is far from an optimal solution. Other ideas may be tried but we will see that the TSP shows a particular difficulty from this point of view.

10.3.1 Concept of approximate algorithms

We are going to define this concept for the case of the TSP, but it can be defined in the same terms in a more general frame for any optimization problem. We call an *instance* of the TSP one case of this problem, that is the datum of a complete graph G weighted at the edges by mapping v with values in $\mathbb{R}^+$. Given an instance of the TSP, we call a *feasible solution* of this instance any Hamilton cycle C of G. An *optimal solution* is a Hamilton cycle C of G for which the total value $v(C)$ is minimal. Let us designate I an instance of the TSP and $R(I)$ the set of all feasible solutions for this instance. We put:

$$v^*(I) = \min_{C \in R(I)} v(C)$$

This quantity is the *optimal value* of instance I. If C^* is an optimal solution of instance I, we have $v^*(I) = v(C^*)$.

An ϵ-*approximate* algorithm for the TSP where ϵ is a real number > 0, is an algorithm which, given an instance I of the problem, returns a feasible solution C which satisfies the following inequality:

$$v(C) - v^*(I) \leq \epsilon v(C)$$

which can also be written by supposing $v^*(I) > 0$, which avoids a trivial case, and $\epsilon < 1$:

$$\frac{v(C)}{v^*(I)} \leq \frac{1}{1-\epsilon}$$

(note that the hypothesis $\epsilon < 1$ is not problematic since, as $v^*(I) > 0$, we always have $v(C) - v^*(I) < v(C)$).

This inequality expresses that the feasible solution C is not further, in relative precision, than ϵ from the optimal value.

Of course, an ϵ-approximate algorithm is only interesting if it is polynomial. We still have to determine for which ϵ such a polynomial algorithm exists. If an algorithm fits the definition for a given ϵ, then it also fits any $\epsilon' \geq \epsilon$. This leads us to define the *approximation threshold* of an optimization problem such as the TSP as the greatest lower bound of the $\epsilon > 0$ for which there exists a polynomial ϵ-approximate algorithm. With regard to what has been noted above concerning ϵ (ϵ can be assumed to be < 1), this approximation threshold is ≤ 1. Only the case of an approximation threshold < 1 teaches us something concerning this problem. A threshold equal to 1 does not teach us anything, and it is the case for the TSP if $\mathbf{P} \neq \mathbf{NP}$, as we are now going to show.

PROPOSITION 10.2. *Except if* $\mathbf{P} = \mathbf{NP}$, *the approximation threshold of the* TSP *is equal to 1.*

Proof. Let us suppose the existence for TSP of a polynomial algorithm, $\mathcal{A}$, which is ϵ-approximate with $\epsilon < 1$. Let us show that such an algorithm would make it possible to solve the problem of the existence of a Hamilton cycle in any graph, a problem which we know is **NP**-complete. So, let $H = (X, F)$ be any graph. Let I be the instance of the TSP defined by the complete graph $G = (X, E)$ considered with the same set X of vertices as H and weighted in the following manner: $v(xy) = 1$ if $xy \in F$, and $v(xy) = \frac{n}{1-\epsilon}$ if $xy \notin F$, where $n = |X|$. Observe that a Hamilton cycle of H corresponds in G to a Hamilton cycle of value n, and, conversely, a Hamilton cycle of value n in G corresponds to a Hamilton cycle of H. Indeed, any edge of G which is not in H has a value $\frac{n}{1-\epsilon} > n$ (in G), and its presence in a cycle of G automatically makes its value greater than n. The problem of the existence in H of a Hamilton cycle therefore can be reduced to that of the existence in G of a Hamilton cycle with a value $\leq n$ (and then in fact equal to n). The application of algorithm $\mathcal{A}$ to G will give a Hamilton cycle C in relation to the optimal value $v^*(I)$ of the instance considered as follows, as it results from the definition of an ϵ-approximate algorithm:

$$v^*(I) \geq v(C)(1 - \epsilon)$$

If C contains an edge which is not in H, we have $v(C) > \frac{n}{1-\epsilon}$, from which we deduce, with the preceding inequality, $v^*(I) > n$. There is therefore no Hamilton cycle with the value n in G. Otherwise, C is a Hamilton cycle of H. We thus see that cycle C given by algorithm $\mathcal{A}$ makes it possible to answer the question of the existence of a Hamilton cycle in H, depending on whether this cycle contains, or does not contain, an edge which is not in H. □

This proposition testifies to the particular difficulty of the TSP, not only in finding an exact solution, as for any **NP**-complete problem, but also in finding an approximate one. Indeed, except if $\mathbf{P} = \mathbf{NP}$, there is no approximate polynomial algorithm useful for the TSP. (Let us recall that an approximation threshold equal to 1 does not yield anything.) This is not the case with all **NP**-complete problems. Indeed, some have an approximation threshold < 1. We are going to see an example with a restriction to the TSP (we will find another example in the exercises).

10.4 Approximation of the metric TSP

The *metric* TSP is defined as the general TSP but with the hypothesis that the values of the edges verify the *triangular inequality*, as is the case with distances in metric spaces. Certain natural applications of the TSP are in this category, such as the one of the robot arm mentioned above. Given an instance I of the metric TSP, defined by a complete graph $G = (X, E)$ weighted by v, we therefore have for any $x, y, z \in X$:

$$v(xz) \leq v(xy) + v(yz)$$

This problem is a particular case of the TSP, but that does not prevent it from also being **NP**-complete. There again a simple strategy like the greedy strategy fails. Nevertheless, this problem is no longer comparable to the case of proposition 10.2. We will indeed give a polynomial ϵ-approximate algorithm with $\epsilon < 1$.

10.4.1 An approximate algorithm

Let us describe step by step a first algorithm. Each of these steps is in itself an algorithm, either already known or described in the usual terms.

Step 1: Construct a minimum spanning tree T of G weighted by v. This step can be achieved with Kruskal's algorithm (Chapter 2).

Step 2: Double each edge of T. The graph thus obtained, U, is Eulerian (Chapter 9). Find a Euler tour D of U, defined by the sequence of its vertices written from any first vertex.

Step 3: "Shorten" the preceding sequence by searching it once again and suppressing any vertex previously visited (except the first when found at the end).

Figures 10.4 and 10.5 give an example of an application of this algorithm.

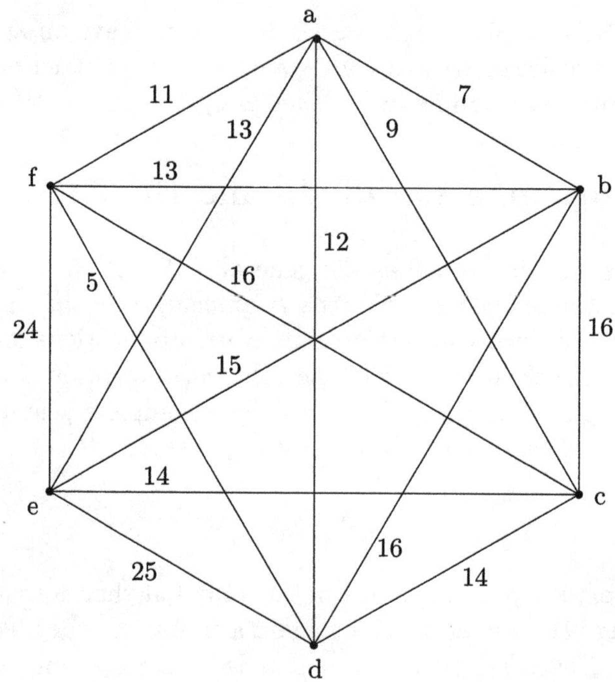

Figure 10.4. *An instance of the metric traveling salesman problem (the values at the edges verify triangular inequality)*

10.4.2 Justification and evaluation

The sequence obtained in step 3 defines a Hamilton cycle of G since each vertex, except the first, appears in it exactly once. Let C be this cycle. What can be said about its total value $v(C)$? A first observation is that any Hamilton cycle of G has a total value greater than or equal to the value $v(T)$ of spanning tree T, because such a cycle deprived of an edge is itself a spanning tree of G, and T is a *minimum* spanning tree. Thus, denoting I the instance considered of the TSP and $v^*(I)$ its optimal value as above, we have:

$$v(T) \leq v^*(I)$$

In addition, the total value $v(D)$ of Euler tour D of U is equal to twice the sum of the values of the edges of T since graph U was obtained by doubling

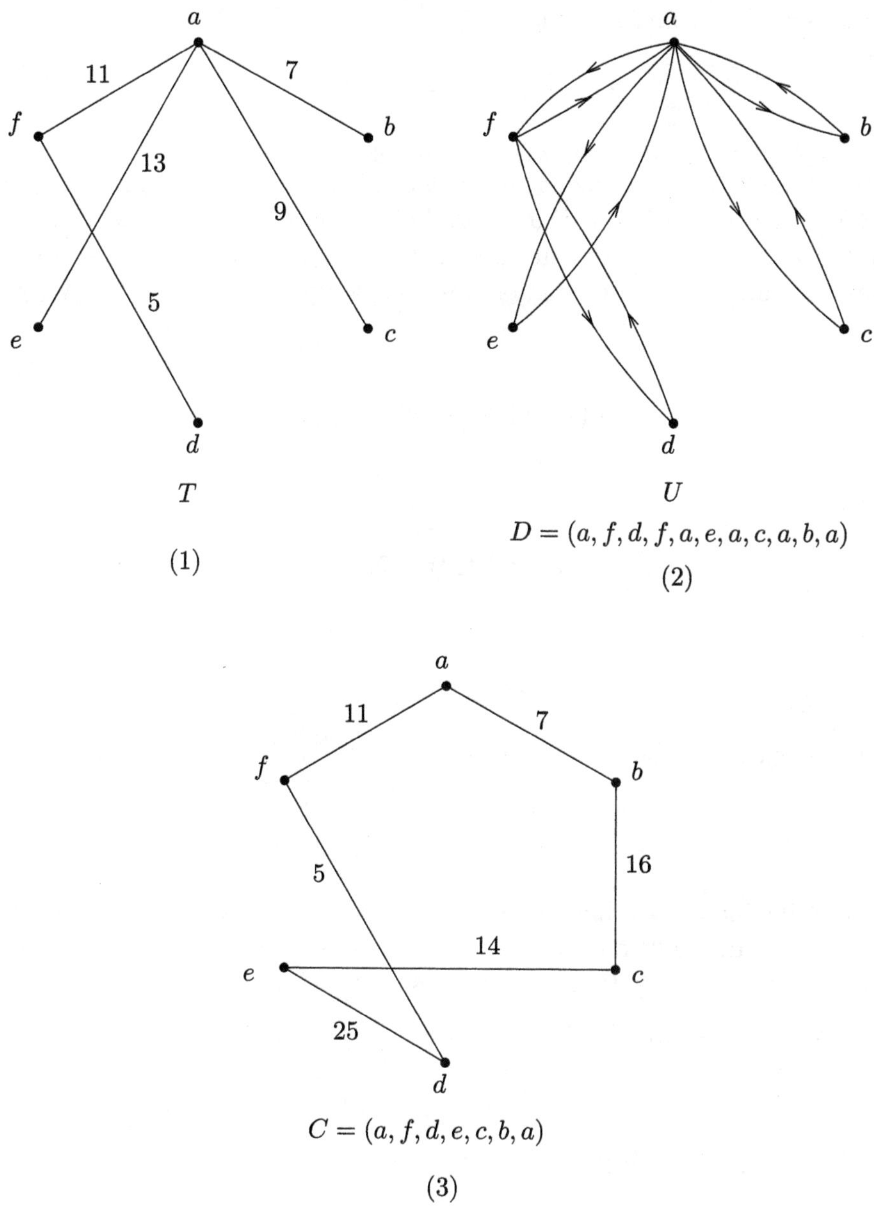

Figure 10.5. *Application of the approximate algorithm to the graph in Figure 10.4. Successively: (1) minimum spanning tree T; (2) graph U obtained by doubling the edges, and Euler tour D (cycle defined on the graph by the arrows of a search); (3) shortening of the preceding sequence and Hamilton cycle C obtained, of value 78*

the edges of T. Therefore, we have:

$$v(D) = 2v(T)$$

The crucial point now is to observe that cycle C, obtained at step 3, has a total value less than or equal to that of D. Indeed, any "shortening" at step 3 comes to replacing in D a subtrail of the form $(x_1, x_2, \dots, x_k)$ by edge x_1x_k (an edge which necessarily exists in G since G is a complete graph). However, thanks to the triangular inequality (possibly repeated), we can write:

$$v(x_1x_k) \leq \sum_{i=1}^{k-1} v(x_ix_{i+1})$$

Thus we have:

$$v(C) \leq v(D) = 2v(T)$$

With the inequality given above for $v(T)$, we obtain:

$$v(C) \leq 2v^*(I)$$

which can also be written:

$$\frac{v(C)}{v^*(I)} \leq \frac{1}{1 - 1/2}$$

This last inequality justifies algorithm $\mathcal{A}$ as being $\frac{1}{2}$-approximate for the metric TSP. In addition, we see that this algorithm is of polynomial complexity by verifying that it is the case for each step.

Thus, we have already shown that the approximation threshold for the metric TSP is $\leq 1/2$.

10.4.3 Amelioration

Let us consider subgraph H of G induced by the vertices of minimum spanning tree T which are, in T, of odd degree. This subgraph is complete, weighted, and has an even number of vertices, since the number of vertices of odd degree of graph T is even. Therefore, it has a perfect matching. Let M be a *minimum* perfect matching of H, that is such that its value $v(M)$ is minimal. Let us then consider graph T augmented by the edges of M, which we call U. By construction, U is connected and has all its vertices of even degree because to each vertex of odd degree of T we have added an edge of matching M. Thus, graph U is Eulerian. Let us then consider a Euler tour D of U defined by the sequence of its vertices, a sequence written from any first vertex. Let us apply to D the same shortening technique as in step 3 of the preceding algorithm. We obtain a sequence of the vertices of U, therefore of G, which defines a Hamilton cycle of G, which we call C. Let us sum up as follows the various steps of this second algorithm.[1]

10.4.4 Christofides' algorithm

Let us describe this algorithm step by step.

Step 1: Construct a minimum spanning tree T of G.

Step 2: Find a minimum perfect matching M in subgraph H of G induced by the vertices which are of odd degree in T.

Step 3: Find a Euler tour D of graph U obtained by adding to T the edges of M, defined by the sequence of its vertices written from any first vertex.

Step 4: "Shorten" sequence D as in step 3 of the preceding algorithm.

Figure 10.6 shows the application of this algorithm to the case in Figure 10.4. We see that a better result is obtained, with a value equal to 66 in place of 78 obtained with the preceding algorithm.

10.4.5 Justification and evaluation

As in the preceding algorithm, C is a Hamilton cycle. Let us try to evaluate $v(C)$. Since C is the shortening of a Euler tour D of U, itself composed of T augmented by M, we have with the triangular inequality:

$$v(C) \leq v(D) = v(U) = v(T) + v(M)$$

Let us now consider C^*, a minimum cycle of G, that is an optimal solution to instance I of the TSP considered with graph G. Let us consider on cycle C^* the succession of the odd degree vertices of T. By associating them by pairs of consecutive vertices on C^*, with the two possible ways following the first

[1]Presentation of this algorithm, and the preceding one, follows the book: *The Traveling Salesman Problem*, Lawler, Lenstra, Rinnooy Kan, Shmoys; Wiley (1985).

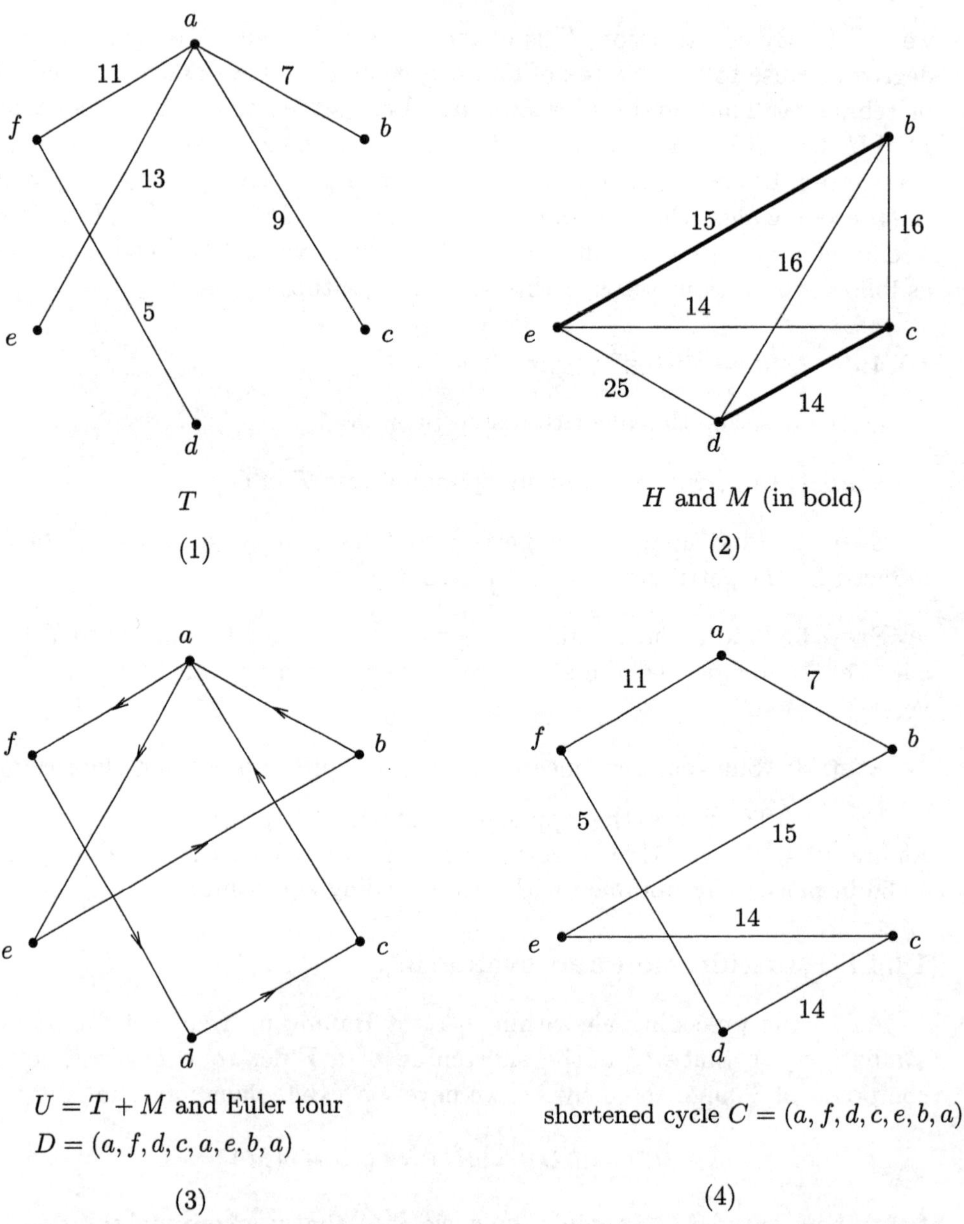

Figure 10.6. *Application of Christofides' algorithm to the graph in Figure 10.4. Successively: (1) minimum spanning tree T; (2) subgraph H induced in the graph by the odd degree vertices of the preceding tree, minimum matching M of this subgraph (in bold lines); (3) graph U and Euler tour D of U defined by the sequence of its vertices from a; (4) shortening of the preceding sequence and Hamilton cycle C obtained, of value 66*

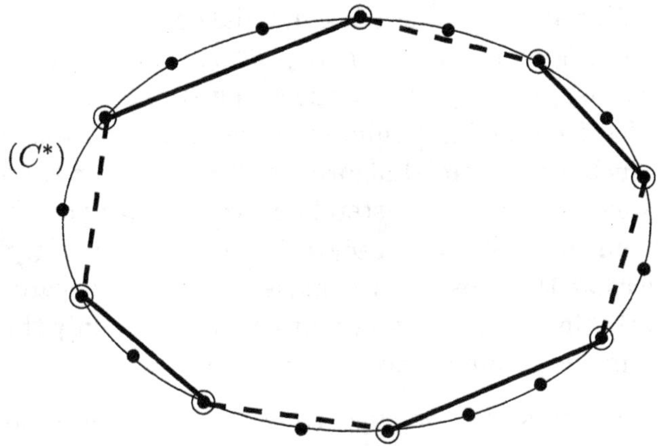

◉ odd degree vertices of T

— edges of M_1

- - edges of M_2

Figure 10.7. *Justification of Christofides' algorithm*

vertex visited, we obtain two matchings M_1 and M_2 of G (see Figure 10.7). The triangular inequality allows us to write:

$$v(M_1) + v(M_2) \leq v(C^*)$$

From this inequality it results that one of the two matchings M_1 or M_2 has a value $\leq \frac{1}{2}v(C^*)$. If for example it is M_1, we have $v(M) \leq v(M_1) \leq \frac{1}{2}v(C^*)$ since M is a minimum matching. In addition we still have the inequality $v(T) \leq v^*(I) = v(C^*)$ (as we have already seen earlier on). Thus:

$$v(C) \leq v(T) + v(M) \leq v(C^*) + \frac{1}{2}v(C^*) = \frac{3}{2}v(C^*)$$

thus:

$$v(C) \leq \frac{3}{2}v^*(I)$$

which can also be written:

$$\frac{v(C)}{v^*(I)} \leq \frac{1}{1 - 1/3}$$

This last inequality justifies the algorithm as being $\frac{1}{3}$-approximate.

In addition, this algorithm has a polynomial complexity. We have already seen this for steps 1 and 4, and for step 3, solved polynomially with the algorithms described in Chapter 9. Concerning step 2, we saw in Chapter 7 a polynomial algorithm giving a minimum matching in a weighted bipartite graph. In fact such a polynomial algorithm also exists for any non-bipartite graph, which solves polynomially step 2. Since we have not given this more general algorithm, we will only apply this step to some simple cases with four odd vertices at the most, which makes it easier to search a minimum matching directly (in the case of four vertices there are only three matchings to compare, as in the example shown in Figure 10.6).

In the end, we can state: *the approximation threshold of the metric TSP is* $\leq 1/3$.

10.4.6 Another approach

We are now going to develop a very different approach, called "heuristic" because it proceeds by local improvements. The principle is to start from a feasible solution, obtained by any appropriate method, here a Hamilton cycle C of complete weighted graph G, and try to improve this solution by modifying C in order to obtain a cycle C' with a strictly lower value.

It is possible to imagine diverse modifications of the cycle. One of the most natural is to replace some of its edges by others such that: 1) we still have a Hamilton cycle, 2) we gain on the values with the exchange of the chosen edges. Let us consider a simple case of such a transformation, which consists of replacing two edges of the cycle by diagonals.

To be specific (see Figure 10.8), taking $(x_1, \ldots, x_n, x_1)$, the sequence of the vertices of C, we replace edges $x_i x_{i+1}$ and $x_j x_{j+1}$ of C, where $1 \leq i < i+1 < j < j+1 \leq n$, with the diagonals $x_i x_j$ and $x_{i+1} x_{j+1}$. Cycle C becomes cycle C' defined by the sequence:

$$\left(x_1, \ldots, x_i, x_j, x_{j-1}, \ldots, x_{i+1}, x_{j+1}, x_{j+2}, \ldots, x_n, x_1\right)$$

and if:

$$v\left(x_i x_j\right) + v\left(x_{i+1} x_{j+1}\right) < v\left(x_i x_{i+1}\right) + v\left(x_j x_{j+1}\right)$$

then we truly have $v(C') < v(C)$. We repeat this modification as long as it is possible.

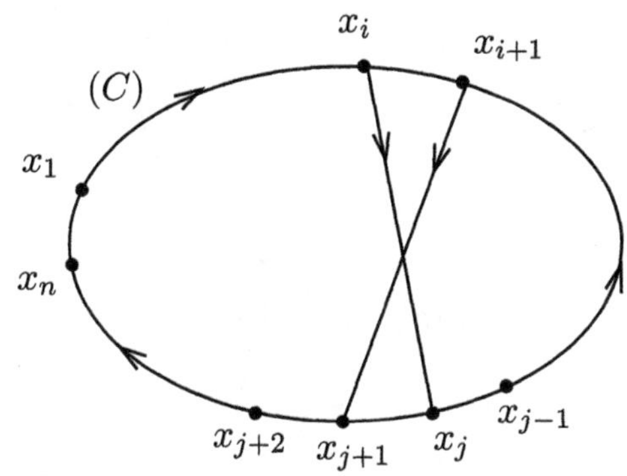

Figure 10.8. *Modification of a Hamilton cycle by a pair of diagonals (the arrows show the search of the modified cycle)*

Figure 10.9 goes back over the example dealt with previously. The Hamilton cycle obtained, by application for as long as possible of this improvement process, has the same value, 66, as the one obtained by Christofides' algorithm (it is in fact the same cycle).

10.4.7 Upper and lower bounds for the optimal value

In practice, it is interesting to repeat this modification process starting from different Hamilton cycles, if possible constructing them in very different ways. However, the question remains of the quality of the solution obtained which, of course, in general has no reason to be optimal. With Christofides' algorithm, for example, we had a "relative error" which, even if it is very large, gives an indication. Generally, an approximate result without precision has no practical value.

In fact, what we need to find is lower and upper bounds for the optimal value $v^*(I)$, where I is the instance considered. For an upper bound, any feasible solution found yields one automatically: for any Hamilton cycle C we indeed have $v^*(I) \leq v(C)$. A lower bound is harder to find. We already saw above such an indication with the consideration of a minimum spanning tree T of G. Indeed, we have: $v(T) \leq v^*(I)$ (remember: cycle C deprived of an edge is itself a spanning tree). We can refine this result by adding to this bound the lowest edge value in the graph (which is less than or equal to the value of the edge removed from C). Let E designate the set of edges of G,

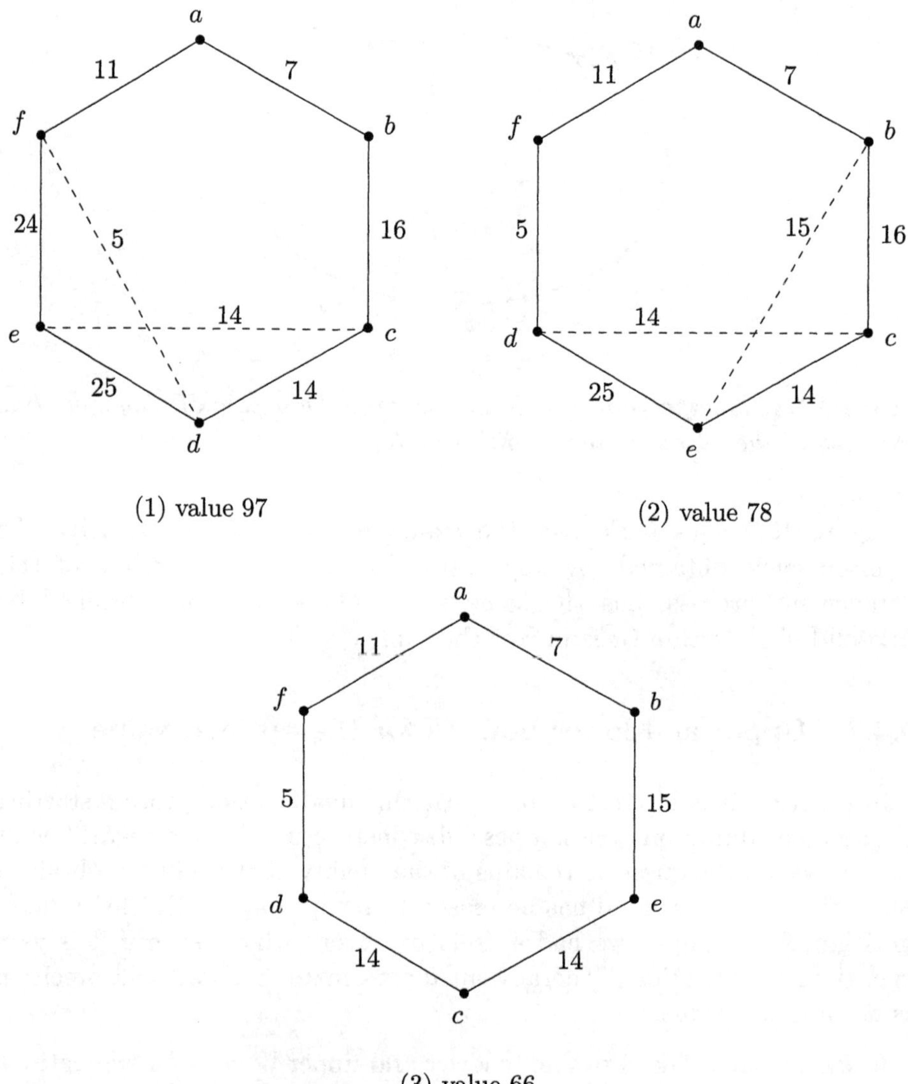

Figure 10.9. *Application to the graph in Figure 10.4 of the improvement by pairs of diagonals. The Hamilton cycle initially considered is the cycle (a, b, c, d, e, f, a), which has the value 97. We choose each time the pair of diagonals which yields the greatest reduction in the value of the cycle. After two modifications (the pair of diagonals considered is indicated each time), we obtain a Hamilton cycle, of value 66, which can no longer be improved by this method*

so we have:

$$v^*(I) \geq v(T) + \min\big(v(e) \mid e \in E\big)$$

In the instance already dealt with (Figure 10.4), this gives $v^*(I) \geq 45 + 5 = 50$.

This technique may be extended to obtain a lower bound greater than or equal to the preceding one, and therefore more interesting. We consider a spanning tree of graph $G - x$ obtained by removing any vertex x from G. Let T_x be such a tree. If C^* is a Hamilton cycle of G, $C^* - x$ is a spanning tree of $G - x$, which implies the inequality $v(C^* - x) \geq v(T_x)$. Let e_1 and e_2 be the edges of C^* incident to x. Let E_x be the set of the edges of G incident to vertex x and let us put the quantity:

$$m_x = \min\big(v(f_1) + v(f_2) \mid f_1, f_2 \in E_x, f_1 \neq f_2\big)$$

We then have:

$$v(e_1) + v(e_2) \geq m_x$$

We deduce from it:

$$v(C^*) = v(C^* - x) + v(e_1) + v(e_2) \geq v(T_x) + m_x$$

Therefore, finally:

$$v^*(I) \geq v(T_x) + m_x$$

If we want to exploit this technique for lower bounding the optimal value to the maximum, it is possible to apply it to all the vertices x of G and to keep the greatest lower bound thus obtained. For the example in Figure 10.4, we show the results in the following table (using the notation used previously).

x	$v(T_x)$	m_x	Lower bound
a	46	7 + 9	62
b	38	7 + 13	58
c	36	9 + 14	59
d	40	5 + 12	57
e	32	13 + 14	59
f	41	5 + 11	57

The largest lower bound obtained is 62. Considering the best solution found earlier on, 66, it finally appears that for this instance of the TSP the optimal value is bounded as follows:

$$62 \leq v^*(I) \leq 66$$

In fact, we have $v^*(I) = 66$ (see exercise 10.4).

10.5 Exercises

10.1. Show that it is not possible to go on a chessboard from the square in the lowest right hand corner to the square in the upper left hand corner by following adjacent squares and visiting all squares at least exactly once (use the colors of the squares).

10.2. *(Dirac theorem)*

Let G be a simple graph such that $n \geq 3$ and $\delta \geq \frac{n}{2}$ (δ being the minimum degree of G). We propose to show that this graph is Hamiltonian.

a) Show that there is a cycle in G.

b) Let us consider a cycle C of G with a maximum number of vertices. Suppose that C is not a Hamilton cycle of G and consider then a vertex x of G which is not in C. Show, using the hypothesis on G, that x is necessarily the neighbor of two vertices of C, themselves neighbors in the cycle. Deduce from this the existence of a cycle containing the vertices of C and vertex x. Conclude.

c) Show that this theorem is a corollary of theorem 10.1 stated (without proof) in this chapter.

+10.3. We consider the greedy algorithm for the TSP. This algorithm consists of doing the following: choose a city of departure, go each time to the nearest city not yet visited, then at the end return to the initial city. Apply this algorithm to the example in Figure 10.4, considering the six possible cases of start vertices. Compare the results obtained.

+10.4. Let us go over the instance of the TSP seen in this chapter (Figure 10.4). Using a *backtracking* (see Chapter 5), give an algorithm to explore all the Hamilton cycles of this graph. Write a program (in the language of your choice) which implements this algorithm. Find, using this program, the optimal value of this instance of the TSP.

10.5. Apply to the metric instance of the TSP on six cites A, B, C, D, E, F, defined by the following table of distances, all approximate resolution methods seen in this chapter. At the end give upper and lower bounds to the optimal solution.

	A	B	C	D	E	F
A	—	5,0	12,5	16,5	25,0	14,0
B	5,0	—	11,5	13,5	24,5	16,0
C	12,5	11,5	—	7,0	13,0	9,5
D	16,5	13,5	7,0	—	15,5	16,5
E	25,0	24,5	13,0	15,5	—	14,0
F	14,0	16,0	9,5	16,5	14,0	—

*10.6. (*An* **NP**-*complete problem with an approximation threshold* < 1)

We are considering the problem of finding in a simple graph $G = (X, E)$ a minimum transversal. Let us recall (Chapter 7) that a transversal L of a graph G is a set of vertices such that any edge of G has at least one endvertex in L. A transversal is called *minimum* if its cardinality is minimal. Let us recall that any transversal has a cardinality greater than or equal to that of any matching, and that if $\nu(G)$ is the matching number of G (greatest cardinality of a matching), then for any transversal L we have $|L| \geq \nu(G)$. This problem of finding a minimum transversal in a graph (expressed as a decision problem) is, as is the TSP, **NP**-complete. Given the following algorithm, in which `H` is a graph, `E(H)` designating the set of its edges, `L` is an auxiliary set of vertices, `x` and `y` are vertices:

```
procedure approx_transv(G);
begin
  H:= G; L:= ∅;
  while E(H) ≠ ∅ loop
    take xy ∈ E(H);
    L:= L ∪ {x,y}; H:= H - {x,y};
  end loop;
end approx_transv;
```

a) Show that the set L obtained is a transversal of G.

b) Show that L is of cardinality $\leq 2\nu(G)$.

c) Deduce from the above that the procedure **approx_transv** is a 1/2-approximate algorithm for the minimum transversal problem.

d) Verify that this algorithm is polynomial (*the approximation threshold of the minimum transversal problem is therefore* $\leq 1/2$).

Chapter 11

Planar Representations

The graphs considered in this chapter are undirected, but in fact directed graphs are covered, since the direction of the edges plays no role in what follows. Likewise, the graphs can be assumed to be simple since loops and multiple edges usually also play no role. The representation of graphs in a plane has been frequently studied, particularly for applications, for example the representation of graphs on a computer screen or the conception of printed circuit boards.

11.1 Planar graphs

Since the study of graphs began, they have been represented in concrete terms by drawing in a plane with points for vertices and lines for edges. The lines used are considered to contain all the mathematical continuity and simplicity properties required ("Jordan curves"). Diverse constraints may be set for this representation. The most natural one is that the lines representing the edges do not intersect except at their endpoints, which defines *planar graphs* (previously defined in Chapter 1). Graphs are not all planar, for example it is quickly realized that the complete graph K_5 is not.

We should make a distinction between a graph G, which is an abstract concept, and its representation on a plane, as defined previously, called *planar embedding*. A planar graph is a graph which allows a planar embedding. There are usually many planar embeddings of a given graph (for example by continuous deformations in the plane of the lines representing the edges).

The connected components (according to the sense of the plane topology) outlined by this representation in the plane are called *faces* of a planar embedding. One of these faces is unbounded; it is called the *unbounded* or *exterior face.* Each face is delimited in the plane by lines which correspond to a closed walk in G. This closed walk is not always a closed trail as we see in Figure 11.1 for one of the faces which is in particular delimited on two
† sides by a common edge. (Which one? Observe the presence of an isthmus in the graph and see exercise 11.1.)

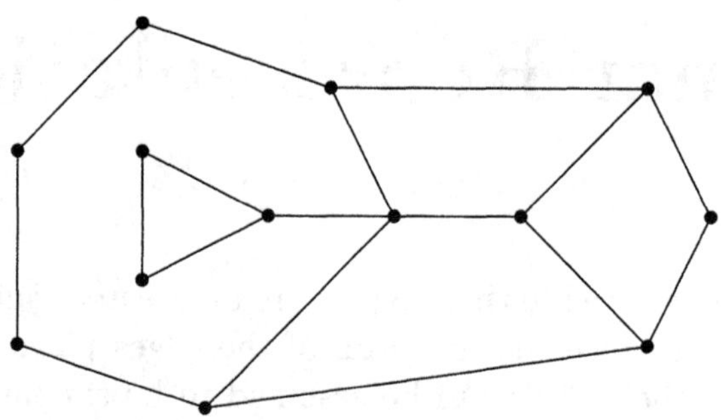

Figure 11.1. *A planar representation of a graph*

11.1.1 Euler's relation

PROPOSITION 11.1. *Given the planar embedding of a connected planar graph G, f the number of faces of this embedding, n the number of vertices, and m the number of edges of G, we have:*

$$n - m + f = 2$$

This relation has in fact been known for a long time, well before graphs, for regular polyhedra (which may be regarded as planar graphs). It can easily be proved by applying inductive reasoning to the number of faces f. An interesting consequence of this relation is that the number of faces f does not depend on the planar embedding under consideration of the graph, since indeed $f = 2 - n + m$. It has many more consequences and we will give a few of them below.

NOTE. Let us recall that we always have $n \geq 1$ for a graph G. Here we also have $f \geq 1$. The particular case $f = 1$ corresponds to the case where connected graph G is a tree. Indeed, since any edge of a tree is an isthmus, in any representation of a tree there is only one face, the unbounded face (it is the same face which is delimited on both sides of each edge; see exercise 11.1). Euler's relation then again gives the relation already known for trees: $m = n - 1$.

COROLLARY 11.1. *If G is a simple planar graph such that $n \geq 3$, then we have:*

$$m \leq 3n - 6$$

Proof. It is enough to prove this equality in the case of a connected graph. Consider a planar embedding of graph G and observe that any face is delimited by a closed walk of G with a length ≥ 3. Indeed, a length equal to 1 or 2 for a face, including the unbounded face, would imply the existence of a loop or a double edge in the graph, which is excluded by hypothesis since the graph is simple. Thus, each face is bounded by at least three edges. Since each edge lines two faces at the most, we deduce the inequality $3f \leq 2m$. By eliminating f between this inequality and Euler's relation, we finally deduce the inequality first stated. □

NOTES. 1) This property of a planar graph is interesting as it is independent from its representations (f is no longer involved).

2) This inequality makes it possible to prove easily that complete graph K_5 is not planar: indeed, we have $n = 5$ and $m = 10$ and thus $m > 3n - 6$, which contradicts the inequality.

3) It should be noted that this inequality is only a necessary condition for any planar graph; it is not a sufficient condition. To verify this, consider bipartite complete graph $K_{3,3}$: we have $n = 6$ and $m = 9$, and therefore $m \leq 3n - 6$, nevertheless it is not a planar graph (see exercise 11.3).

4) The case of the equality $m = 3n - 6$ corresponds to what is called the *triangulations of the plane*, that is the planar embeddings in which any face is delimited by a triangle (cycle of length 3). Simple graphs thus represented are *maximal* planar graphs, meaning that any edge added cancels the property of planarity.

The following corollary has played a fundamental role in the proof of the famous four-color theorem.

COROLLARY 11.2. *If graph G is simple and planar, then it has a vertex of degree ≤ 5.*

Proof. For $n = 1$ or $n = 2$ this property is trivially true. Let us therefore suppose that $n \geq 3$ and let us make δ_G the minimum degree of graph G considered. We have to show that $\delta_G \leq 5$. We have, by lower bounding each term of the sum of the degree of the graph by δ_G:

$$n\delta_G \leq \sum_{x \in X} d_G(x) = 2m$$

and thus, with the inequality from corollary 11.1:

$$n\delta_G \leq 2 \times (3n - 6) = 6n - 12$$

and finally:

$$\delta_G \leq 6 - \frac{12}{n} < 6$$

which gives the inequality desired. □

11.1.2 Characterization of planar graphs

There are several well-known necessary and sufficient conditions for a graph to be planar. The following one is one of the first to have been given (in the 1930s) and has a great impact on the theory. It is the condition of *excluded configurations.*

THEOREM 11.1 (Kuratowski). *A graph is planar if and only if it does not contain, as a subgraph, a subdivision of K_5 or of $K_{3,3}$.*

We call *subdivisions of a graph H* any graph obtained by replacing some edges of G by paths with lengths ≥ 2, these paths having no common intermediary vertices. In concrete terms, it is as if we added to some of the edges of H one or more vertices of degree 2 (see Figure 11.2). With regard to this theorem, graphs K_5 and $K_{3,3}$ play a particular role: these are (to more or less one operation of subdivision) the configurations whose presence makes a graph non-planar. Conversely, their non-presence implies that the graph is planar.

It is quite easy to show the necessary condition of Kuratowski's theorem, knowing that K_5 and $K_{3,3}$ are not planar. Let us first observe the following point, which is easy to verify:

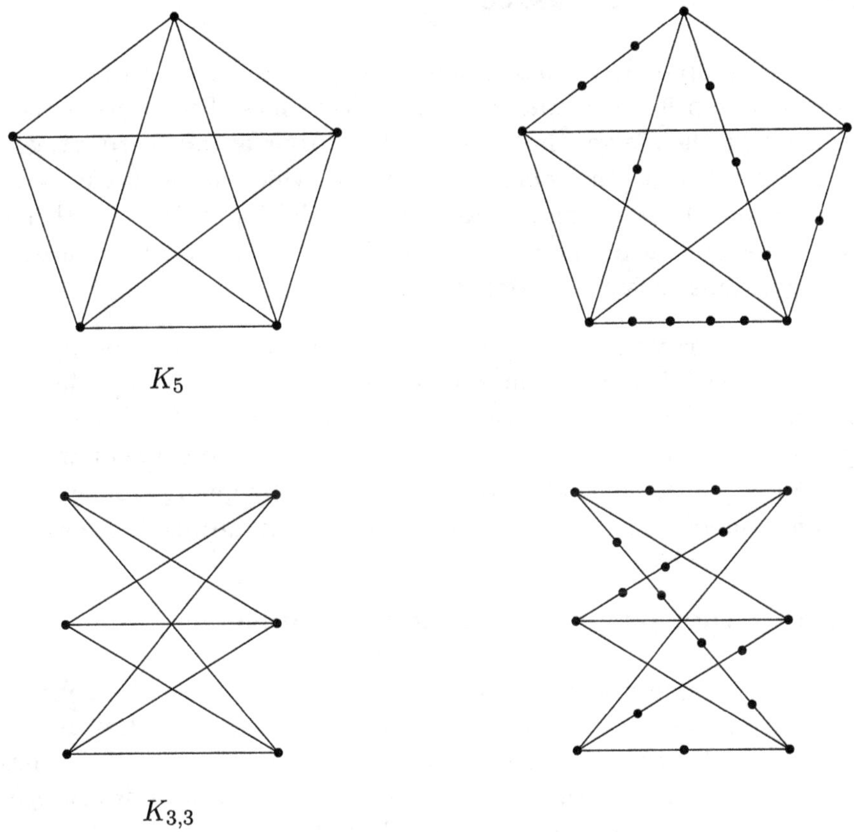

Figure 11.2. *At the top: K_5 and one of its subdivisions, at the bottom: $K_{3,3}$ and one of its subdivisions*

- *If a graph is planar, all its subgraphs are planar.*
- *If a graph is not planar, none of its subdivisions is planar.*

Therefore, if a graph contains as a subgraph a subdivision of K_5 or of $K_{3,3}$, it cannot be planar since it contains a non-planar subgraph.

The sufficient condition of Kuratowski's theorem is harder to show and we will accept it.

11.1.3 Algorithmic aspect

Planar graphs have interesting applications, for example for the conception of printed circuit boards in electronics. People were therefore interested in their algorithmic recognition from a very early stage. The algorithm belongs to the complexity class **P**, which means it can be solved by a polynomial algorithm, as was proven in the 1960s. On the other hand, it was more difficult to find a *linear* complexity algorithm. There are several such algorithms today, but none is truly simple.

The planarity property typically illustrates the concept of a *well-characterized* property in the sense used in complexity theory (see Appendix B). Indeed, when the question is to know if a given graph G is planar, the answer can be *certified* for both possible answers: if it is *yes*, by giving a planar embedding of G, if it is *no*, by giving a subgraph of G which is a subdivision of K_5 or $K_{3,3}$ (following Kuratowski's theorem).

11.1.4 Other properties of planar graphs

There are many other properties of planar graphs. The subject was studied in depth from the origin of graph theory, in particular because of the four-color theorem. Let us quote Fary's theorem as an example: *a planar graph can be embedded in the plane such that all edges are straight-line segments (which do not intersect outside of their endpoints).* Also, Tutte's theorem: *a 4-connected planar graph is Hamiltonian.* The question of planar graphs has also been extended to other surfaces of the plane, for example the torus (orientable surface with a hole), for which we have the equivalent of the four-color theorem with seven colors. The general problem has now been completely resolved.

11.2 Other graph representations

For want of being able to represent a graph with no crossing of edges, other representations were sought in order to avoid the constraint of a representation in a plane with no crossing of edges.

11.2.1 Minimum crossing number

The first idea, naturally, is to reduce to a minimum the number of crossings of edges (outside of the endvertices). The planar case corresponds to the case of a representation without crossings. Let cr(G) be the minimum number of edges crossing in a planar embedding of a graph G. The determining of this parameter is algorithmically difficult; to be more specific, it is an **NP**-complete problem (in the form of a decision problem, that is, given an integer k, do we have cr(G) $\leq k$?). Thus, there is little hope for a good general algorithmic solution. Nevertheless, we have this result for the complete graph:[1]

$$\mathrm{cr}\left(K_n\right) \leq \frac{1}{4}\left\lfloor\frac{n}{2}\right\rfloor\left\lfloor\frac{n-1}{2}\right\rfloor\left\lfloor\frac{n-2}{2}\right\rfloor\left\lfloor\frac{n-3}{2}\right\rfloor$$

EXAMPLE. For $n = 5$, we have:

$$\mathrm{cr}\left(K_5\right) \leq \frac{1}{4} \times 2 \times 2 \times 1 \times 1 = 1$$

Since K_5 is not planar, we have $\mathrm{cr}(K_5) > 0$ and therefore $\mathrm{cr}(K_5) = 1$.

11.2.2 Thickness

Another way of doing this is to decompose a graph into planar subgraphs. It is interesting, for example, to decompose an electronic circuit into subcircuits, each printable on a plate. We want to limit the number of plates, that is the number of subgraphs entering into the decomposition. Since each edge must be in one of the subgraphs and isolated vertices play no role in the planarity, the subgraphs may be defined as spanning subgraphs.

We thus define the *thickness* $t(G)$ of a graph $G = (X, E)$ as the lowest integer k for which there exists a partition of set E; let this be $(E_1, E_2, \ldots, E_k)$, such that $G_i = (X, E_i)$ is planar for $i = 1, 2, \ldots, k$. The determination of the thickness is again an **NP**-complete problem. Nevertheless, we have the following specific case.

[1] Let us recall the notation: $\lfloor x \rfloor$ to designate the greatest integer $\leq x$ and $\lceil x \rceil$ to designate the least integer $\geq x$.

PROPOSITION 11.2. *The thickness $t(G)$ of a graph G such that $n \geq 3$ verifies the inequality:*

$$t(G) \geq \left\lceil \frac{m}{3n-6} \right\rceil$$

Proof. Use the definition of thickness. By applying the inequality in corollary 11.1 to each G_i, putting $m_i = |E_i|$, we have for $i = 1, 2, \ldots, k$:

$$m_i \leq 3n - 6$$

and by performing the sum:

$$m = \sum_{i=1}^{k} m_i \leq k(3n-6)$$

thus:

$$k \geq \left\lceil \frac{m}{3n-6} \right\rceil$$

which implies the stated inequality. □

In particular, it is interesting to apply this inequality to complete graph K_n (see exercise 11.5).

11.3 Exercises

+11.1. a) Given a planar embedding of a connected graph G, show, by the absurd, that if an edge e is an isthmus then it will delimit the same face on both sides.

b) By applying the preceding property, show that in any planar embedding of a tree there is only one face.

*11.2. Show that any graph can be represented in three-dimensional Euclidian space by points for the vertices and lines for the edges in such a way that two lines do not intersect outside their endpoints (in other words, the equivalent in space of the planar embedding problem is trivial).

+11.3. a) Show by applying Euler's relation that if a simple connected graph G such that $n \geq 3$ is planar and without triangles (without length 3 cycles), then it verifies the inequality:

$$m \leq 2n - 4$$

(reason as for the proof of corollary 11.1).

b) Deduce from the preceding a proof that complete bipartite graph $K_{3,3}$ is not planar.

11.4. Let us consider the following graph (well known by the name of *Petersen's graph*):

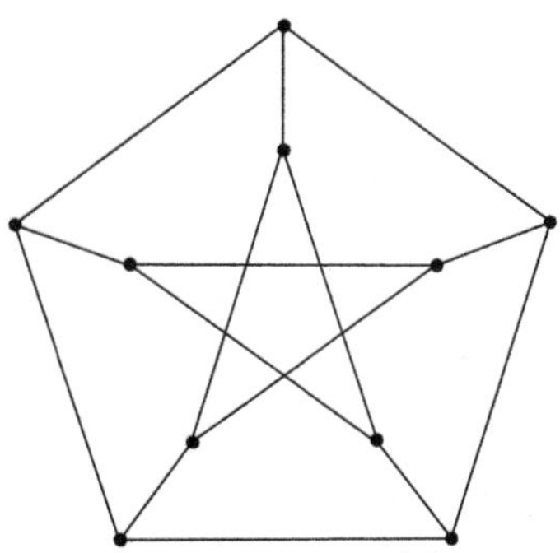

a) Find in this (non-planar) graph a subdivision of $K_{3,3}$.

b) Why is there no subdivision of K_5 in this graph?

*11.5. a) Applying to the complete graph K_n the inequality on thickness shown in this chapter, prove the inequality:

$$t(K_n) \geq \left\lceil \frac{n(n-1)}{6n-12} \right\rceil = \left\lfloor \frac{n+7}{6} \right\rfloor$$

b) Show that we have the equality until $n = 8$ (in particular we will have to show that $t(K_8) \leq 2$, that is find a decomposition of K_8 into two planar spanning subgraphs, which can be done directly).

Chapter 12

Problems with Comments

12.1 Problem 1: A proof of k-connectivity

12.1.1 Problem

Given two integers n and k such that $0 < k < n$ and k is *even*, the simple (undirected) graph $H_{k,n}$ is defined in the following way:

- $X = \{0, 1, \ldots, n-1\}$ is the set of vertices,
- for $i, j \in X$, ij is an edge of $H_{k,n}$ if and only if we have $j = i \pm r$, where $1 \leq r \leq \frac{k}{2}$ with the operation $\pm$ taken *modulo* n. For instance, with $n = 8$, 7+1 is equal to 0, 7+2 is equal to 1, or even 0-1 is equal to 7, etc.

1) What is the number of edges of $H_{k,n}$?

2) Verify that the graph $H_{k,n}$ is connected.

3) A is a set of vertices of $H_{k,n}$ such that the subgraph $H_{k,n} - A$ is not connected. We intend to show that $|A| \geq k$. Reasoning by contradiction, suppose we have $|A| < k$. Then, let i and j be two vertices which belong to different connected components of $H_{k,n} - A$. We set:

$$S = \{i, i+1, \ldots, j-1, j\}$$

and:

$$T = \{j, j+1, \ldots, i-1, i\}$$

In these expressions of S and T, the indicated operations are always to be considered *modulo n*. As $|A| < k$ and as the sets $A \cap S$ and $A \cap T$ are disjoint, we have $|A \cap S|$ or $|A \cap T| < \frac{k}{2}$. Suppose $|A \cap S| < \frac{k}{2}$.

a) Show that there is a sequence of vertices from i to j in $S \setminus A$ such that the difference of two consecutive sequences is at most equal to $\frac{k}{2}$.

b) Deduce from the preceding that the graph $H_{k,n}$ is k-connected.

4) Given two integers as above, we define as $f(n,k)$ the least number of edges that can have a simple graph on n vertices which is k-connected.

a) Show that we have in a general way, whether integer k is *even or odd*:

$$f(k,n) \geq \left\lceil \frac{kn}{2} \right\rceil$$

($\lceil x \rceil$ denotes the least integer $\geq x$)

b) Show that we have equality when integer k is even.

5) Consider the same problem with *k-edge-connected* in place of *k-connected*.

12.1.2 Comments

In this problem, we determine the minimum edges of a simple k-connected graph on n vertices, in the case where n is even. It is a solution to the general problem of communication networks, defined in Chapter 2, in the particular case of a complete graph with a value equal to 1 on each edge.

We define a family of graphs achieving this minimum, equal to $\lceil \frac{kn}{2} \rceil$. The only point which is a little difficult is proof of the k-connectivity. The method proposed in exercise 2.18 (Chapter 2), for the case $n = 8, k = 4$, by applying Menger's theorem, does not lead directly to a generalization.

It is interesting to note that this minimum of edges is the same with the hypothesis *k-edge-connected* instead of *k-connected* (question 5).

The case where k is odd can be handled in the same way; the considered graph $H_{k,n}$ is similar, but a little more complicated to define.

12.2 Problem 2: An application to compiler theory

12.2.1 Problem

Let $G = (X, A)$ be a strict digraph which has a *root* r, which means that for each vertex v of G there is a path from r to v. Given two vertices v and u of G, we say that vertex v *dominates* vertex u, or that v is a *dominator* of u, if each path from r to u goes through v. In particular, each vertex is a dominator of itself. For each vertex u of G, $D(u)$ is the set of the dominators of u. We say that an arc (v, u) of G is a *return arc* if vertex u dominates vertex v. R is the set of return arcs of G. The digraph G is called *reducible* if the spanning subdigraph $H = (X, A \setminus R)$, induced by the non-return arcs, is without circuits.

To illustrate this problem, we will consider the digraph in Figure 12.1.

1) Show the following equality for each $u \in X$:

$$D(u) = \{u\} \cup \left(\bigcap_{(v,u) \in A \setminus R} D(v) \right)$$

2) Determine the return arcs of the digraph in Figure 12.1. Is this digraph reducible?

3) We return to a general digraph G, assumed to be reducible. Suppose that you know an acyclic numbering of $H = (X, A \setminus R)$, and show that it is possible to determine, step by step following the acyclic numbering, the sets $D(u)$ for $u \in X$.

4) We consider a depth-first search of G starting at vertex r. Show that the numbering in postvisit of the vertices of G, in decreasing order from n to 1 (where $n = |X|$), is an acyclic numbering of H.

5) Applying the two preceding questions, determine the sets $D(u)$ of the digraph in Figure 12.1.

12.2.2 Comments

This problem is extracted from an application to the compiler theory, which concerns analysis of loops of a program in the code optimization phase.[1]

[1]This application is described completely in the book *Principles of Compiler Design*, from A. V. Aho, J. D. Ullman, Addison-Wesley (1977).

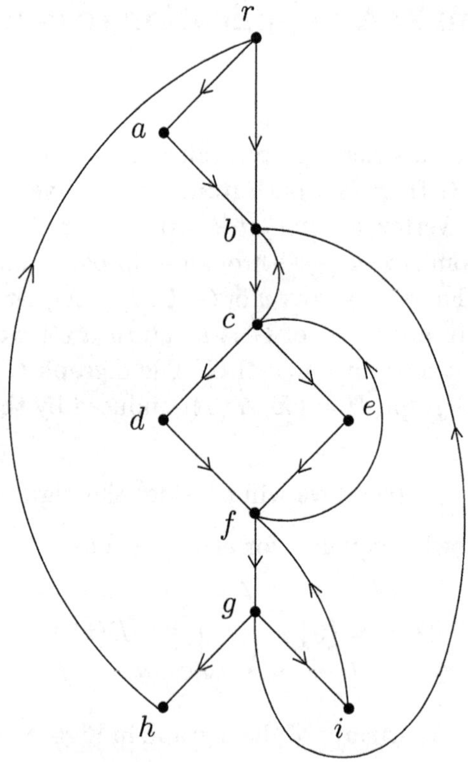

Figure 12.1. *Problem 2*

The proof of the equality in question 1 is not straightforward. Be aware of the case $u = r$, where $D(r) = \{r\}$ because vertex r has no other dominator than itself.

What follows in the problem describes the elements of an algorithm for determining the domination sets of a reducible digraph. The solutions are not trivial, but there is no particular difficulty. Note in question 4 that the depth-first search is made on digraph G, while the acyclic numbering is obtained for spanning subdigraph H.

To enable verification of the results obtained, we give below the domination sets of the vertices of the given digraph (which is reducible), sets which are asked for in question 5.

u	$D(u)$
r	$\{r\}$
a	$\{r, a\}$
b	$\{r, b\}$
c	$\{r, b, c\}$
d	$\{r, b, c, d\}$
e	$\{r, b, c, e\}$
f	$\{r, b, c, f\}$
g	$\{r, b, c, f, g\}$
h	$\{r, b, c, f, g, h\}$
i	$\{r, b, c, f, g, i\}$

12.3 Problem 3: Kernel of a digraph

12.3.1 Problem

Let $G = (X, A)$ be a strict digraph. The *kernel* of G is a set of vertices N of G which verifies the two following properties:

1. N is a *stable set*, that is there is no arc (x, y) of G with $x, y \in N$,

2. N is *dominating*, that is for each vertex $x \notin N$ there exists an arc (x, y) of G with $y \in N$.

We suppose that digraph G is without *odd* circuits (an odd circuit is a circuit of *odd* length), and we intend to show the existence of a kernel in G.

1) Suppose first that digraph G is strongly connected.

a) Given any fixed vertex x_0 of G, show that for each vertex y of G the paths from x_0 to y have lengths which are all of the same parity (that is all even or all odd). We can accept the following technical result: in a digraph, we can extract an odd circuit from any closed directed walk which is of odd length.

b) Show that the vertices of G which are ends of even directed walks beginning at x_0 constitute a kernel of G (this property is also true for the vertices which are ends of *odd* directed walks).

2) We now return to any digraph G.

a) Show that in G there is a strongly connected component, called a *sink*, which has no exiting arcs (each arc whose tail is in this component has its head in this component). We can consider the digraph of the strongly connected components; remember that this digraph is without circuits (see exercise 4.8 in Chapter 4).

b) Let C_1 be a sink strongly connected component of G. We know, from the first question, that C_1 has a kernel, say N_1. We consider the subdigraph G_2 of G which is obtained by removing C_1 of G and also the vertices of G which are the tail of an arc whose head is in N_1. Suppose that G_2 has a kernel, say N_2. Show that $N_1 \cup N_2$ is a kernel of G.

c) Show, by induction on its number of vertices, that if G has no odd circuits then it has a kernel.

3) Apply the preceding to the digraph in Figure 12.2.

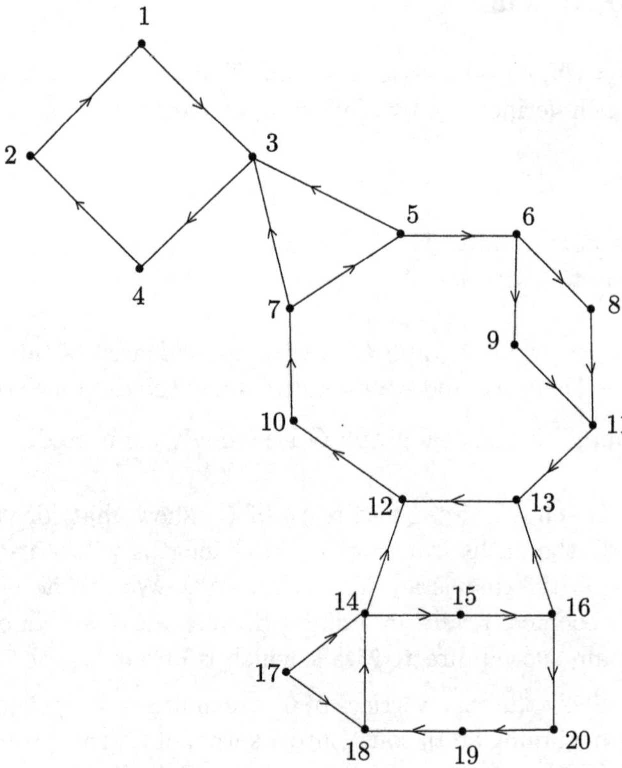

Figure 12.2. *Problem 3*

12.3.2 Comments

The concept of the kernel of a digraph is linked to the search for a winning strategy for a two-player game. We saw in Chapter 5 the example of Nim's game, and exercise 5.4 proposes the determination of a winning strategy for this game using a kernel of some associated digraph.

Our problem here studies a class of digraphs for which there is a kernel. We can deduce from this an algorithm for constructing a kernel in a digraph without odd circuits. This construction needs an algorithm for the determination of the strongly connected components. There are indeed good algorithms for this (not developed in this book).

12.4 Problem 4: Perfect matching in a regular bipartite graph

12.4.1 Problem

Let $G = (X, Y, E)$, where $X = \{x_1, \dots, x_2\}$ and $Y = \{y_1, \dots, y_2\}$, be a (undirected) connected bipartite graph which is k-regular ($k \geq 2$). Graph G may have multiple edges. Suppose a point z to be moving in G and transforming G according to the following rules:

(i) First located at vertex x_1, point z moves along any edge with endvertices x_1 and y_j to y_j, this edge then being removed.

(ii) If z is at y_j, coming from x_i, then it moves to the first vertex $x_l \neq x_i$ (according to the order of the indices), which is a neighbor of y_j, along an edge with endvertices x_l and y_j, this edge then being doubled (that is replaced by two edges with endvertices x_l and y_j).

(iii) If z is at x_i, coming from y_j, then it moves over to the first vertex $y_h \neq y_j$ which is a neighbor of x_i (according to the order of the indices), along an edge with endvertices x_i and y_h, this edge then being removed.

1) Show that this sequence of moves necessarily terminates when point z comes back to vertex x_1, and x_1 is joined to one of the vertices y_j by k parallel edges.

2) Show that the condition described in the preceding question necessarily occurs after a finite number of moves of z.

3) Deduce, from the preceding, an algorithm for finding a perfect matching in a connected bipartite graph which is k-regular.

12.4.2 Comments

We know that a regular bipartite graph accepts a perfect matching (see the "Marriage lemma" in Chapter 7). This problem describes an interesting method for finding such a matching, using only the operations of deletion and joining of edges, as we could do on a blackboard with an erazer and a piece of chalk.[2]

Question 2, which establishes the finiteness of the algorithm, is a little more difficult. Reasoning by contradiction, that is by supposing that the sequence of moves does not terminate, consider the edges of G which are parallel to edges which are searched an infinite number of times. We can show that a vertex of G cannot have more than two such incident edges, and we deduce the existence in G of a cycle which is composed of these edges. This fact implies that some edges must be searched infinitely often *in the same direction*, which is not possible.

12.5 Problem 5: Birkhoff-Von Neumann's theorem

12.5.1 Problem

A square matrix whose entries are real numbers ≥ 0 is said to be *bistochastic* if for each row and each column the sum of the terms is equal to 1. A *permutation* matrix is a square matrix whose terms are 0 or 1 and which has exactly one 1 in each row and one 1 in each column. It is a particular bistochastic matrix. Remember that a *diagonal* of a square matrix, of order n, is a set of n terms of the matrix such that there is one term per row and one term per column.

Let $A = (a_{ij})$, where $1 \leq i, j \leq n$, be a square matrix of order n which is bistochastic.

1) We associate with matrix A the bipartite graph $G = (X, Y, E)$ defined as follows: $X = \{x_1, \ldots, x_n\}$, $Y = \{y_1, \ldots, y_n\}$, and $x_iy_j \in E$ if an only if

[2]This problem is taken from Lovász [9].

$a_{ij} > 0$. By applying Hall's theorem (theorem 7.2 in Chapter 7), show that there is a diagonal in A which is formed of non-zero terms of A.

2) Using the preceding diagonal, show that there exists a permutation matrix P_1 and a real number $\lambda_1 > 0$ such that the matrix $A_1 = A - \lambda_1 P_1$ is bistochastic, up to a positive coefficient (that is there is a real number $\mu > 0$ such that matrix μA_1 is bistochastic).

3) Continue the preceding process in order to show that each bistochastic matrix A is of the form:

$$A = \lambda_1 P_1 + \cdots + \lambda_k P_k$$

where P_i, for $i = 1, \ldots, k$, are permutation matrices, and λ_i are real numbers > 0 such that $\sum_{i=1}^{k} \lambda_i = 1$. In other words, *a bistochastic matrix is a barycenter of permutation matrices.*

4) Application: in a society, an equal number n of men, $m_1, \ldots, m_n$, and of women, $w_1, \ldots, w_n$, spend their lives as couples. For each couple (m_i, w_j) the proportion of life spent together is defined by a_{ij}, with $0 \leq a_{ij} \leq 1$ (it is supposed that all have the same life expectancy). Therefore, we must have for each i, $\sum_{j=1}^{n} a_{ij} = 1$ and for each j, $\sum_{i=1}^{n} a_{ij} = 1$. Thus, the square matrix of order n, $A = (a_{ij})$ is bistochastic. In addition, we give for each couple (m_i, w_j) a "satisfaction coefficient", c_{ij}, of their life in common, which when multiplied by their proportion of life in common gives the "life satisfaction". Thus, a couple (m_i, w_j) has a satisfaction with their life in common equal to $c_{ij} a_{ij}$. We are trying to make the total satisfaction maximum. More specifically, we are looking for some a_{ij}, that is couples lasting, so that $\sum_{ij} c_{ij} a_{ij}$ is maximum. Show that this maximum will be reached in the case of couples stable all their life, that is such that for each i there is a j such that $a_{ij} = 1$.

12.5.2 Comments

The simplicity of the proof of Birkhoff-Von Neumann's theorem proposed in this problem comes from the use of Hall's theorem. It is a nice example of a simple proof of a mathematical theorem thanks to combinatorics.[3]

Question 4 proposes an amusing application of this theorem. The result may appear unpleasant for some ..., but, remember, we can equally observe

[3] From Berge [1], see also: Lawler, *Combinatorial Optimization*, Holt, Rinehart, Winston (1976).

that a *minimum* is also attained in the same condition for couples which are stable all their life!

12.6 Problem 6: Matchings and tilings

12.6.1 Problem

We consider bounded regions of a plane consisting of unit squares. These regions, called *polyominoes*, are supposed connected and even simply connected, that is their boundary is a simple closed polygonal line. We study *tilings* of such regions by non-overlapping *dominoes*, which are 2×1 or 1×2 rectangles. A polyomino which accepts such tiling is said to be *tilable*. We suppose that the squares of the plane are colored black and white, like a chessboard. Figure 12.3(a) gives an example of a tilable polyomino (dominoes of the tiling are surrounded in bold).

1) Show that if a polyomino is tilable then it contains an equal number of white squares and of black squares. We will call a polyomino with this property *balanced*.

2) Given a polyomino P, we associate the (undirected) graph, simple and connected, defined in the following way: to each square of P is associated a vertex of G, and the edges of G correspond to the pairs of squares of P which are adjacent (horizontally or vertically).

a) Show that G is a bipartite graph.

b) Show that polyomino P is tilable if and only if graph G has a perfect matching.

c) Consider the polyomino in Figure 12.3(b). It is balanced, but show that it is not tilable by applying Hall's theorem (theorem 7.2 in Chapter 7) to its associated graph.

3) A polyomino is said to be *trapezoidal* if it is made up of a stack of dominoes in lines which are decreasing in length from bottom to top. An example is given in Figure 12.3(c).

a) Show that a trapezoidal polyomino is tilable if and only if it is balanced. Reason by induction on the number of squares of the polyomino (a number which is necessarily even). Show, in particular, that if a polyomino does not have a pair of

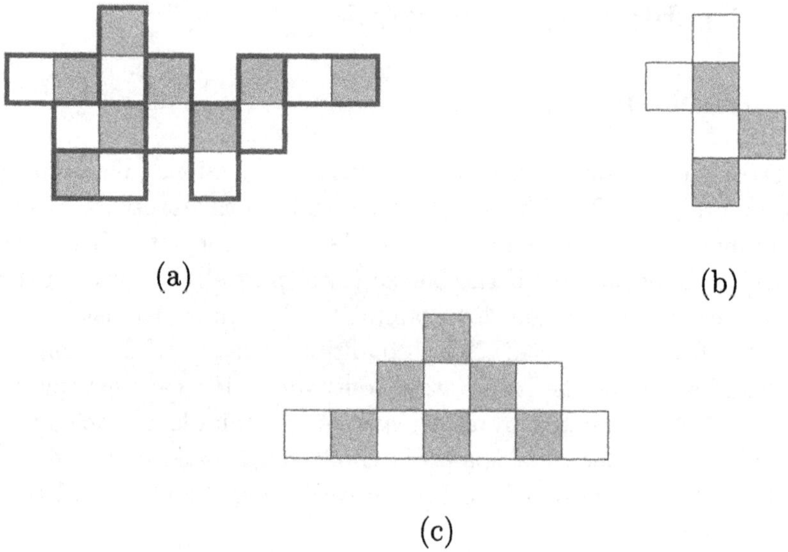

Figure 12.3. *Problem 6*

adjacent squares that we can remove, and the polyomino remains trapezoidal, then the polyomino has the form of a "double staircase", which means that it is not balanced.

b) Deduce, from the preceding reasoning, an algorithm for finding a tiling of a trapezoidal polyomino. What is its complexity?

12.6.2 Comments

Tilings studied in this problem are limited to domino tilings of polyominoes. Tiling problems are much more general, but this particular case is already rich, and, moreover, is in fact relevant to matchings in bipartite graphs, which is our motivation here.[4]

[4]The reader who wants a deeper study of this subject can consult the following reference: J.C. Fournier, "Pavage des figures planes sans trous par des dominos: Fondement graphique de l'algorithme de Thurston, parallélisation, unicité et décomposition", *Theoretical Computer Science*, vol. 159, 1996, 105–128.

12.7 Problem 7: Strip mining

12.7.1 Problem

A strip mine is being exploited by block excavations. These are set as indicated in Figure 12.4, which is the case study in this problem. Each block bears a number (between parentheses at the top to the left). The excavation of a block is only possible if the blocks (even partially) located just above this block are also excavated (for example block number 1 must be removed for number 6 to be removed). This condition is repeated from one row to another and we call it the "excavation constraint". For each extracted block there is a profit indicated by an integer in the middle of each block (for example block number 1 yields a profit equal to 2). This negative or positive value depends essentially on the cost of excavating the block and the value of its ore.

We want to define a set of excavating blocks which respects the excavating constraint and such that the sum of profits obtained by excavating these blocks is the greatest possible. Such a set is said to be *optimum*.

Given a block b, we denote by $p(b)$ the profit obtained by excavation of b. We denote by A the set of blocks b such that $p(b) > 0$ and by B the set of blocks b such that $p(b) < 0$. We model the problem by associating with it the transportation network R which is defined as follows:

- its vertices are the blocks, plus two vertices s (source) and t (sink),
- if b and b' are two blocks such that block b' lies in the mine at the level just above that of b, and b' overlaps b, totally or partially, there is in R an arc $a = (b, b')$, with capacity ∞,
- there is an arc (s, b) of capacity $p(b)$ for each block $b \in A$, and an arc (b, t) with capacity $-p(b)$ for each block $b \in B$.

1) Draw the network which is associated with Figure 12.4.

2) Show that a set C of blocks respects the excavating constraint if and only if in the network R the capacity of the cut $K = \omega^+(C \cup \{s\})$ is finite.

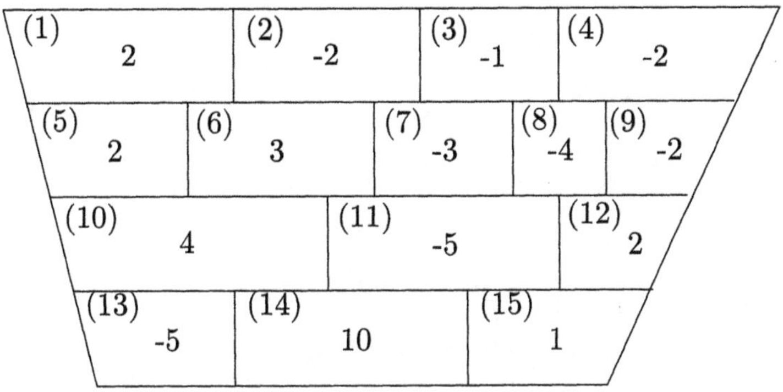

Figure 12.4. *Problem 7*

3) Show that we have, in the case of the preceding question, the following expression for the capacity of the considered cut:

$$c(K) = \sum_{b \in A \setminus C} p(b) - \sum_{b \in B \cap C} p(b)$$

Deduce:

$$c(K) = \sum_{b \in A} p(b) - \sum_{b \in C} p(b)$$

4) Show that a set C of blocks is optimum if and only if the preceding cut K is minimum.

5) Find such an optimum set of blocks by applying the max-flow min-cut algorithm to network R.

6) Determine for block 14 a profit (> 10) from which this block can belong to a set of blocks which is optimum. Determine then such a set of blocks.

12.7.2 Comments

The application developed in this problem presents the particular interest that it uses the Ford-Fulkerson theorem and algorithm for a *minimum cut* instead of for a *maximum flow*, as is most frequently the case.[5]

There is no particular difficulty in this problem.

[5]The reader can find a general presentation of this application in the book: *Linear Programming*, V. Chvátal, Freeman (1983).

Appendix A

Expression of Algorithms

The formal expression of algorithms is a delicate question and the solutions given in the abundant literature on algorithms are diverse and, it should be said, not always appropriate. There are in fact opposing objectives: the wish to remain close to the natural human language and the relative necessity of being able to translate into a real programming language, in order to validate the algorithm expressed. It is of course possible to limit ourselves, in certain cases, to an expression in natural language. This renders the *principle* of the algorithm rather than the algorithm itself. A more in-depth study requires a more formalized expression, in particular in order to analyze complexity (see Appendix B).

It is necessary to apply the model of "structured programming", much used today by most commonly used languages such as C for example. In fact the ADA language is a much better reference for programming algorithms; it is cleaner, more conceptual, but unfortunately has not (yet?) met with the favor of the public. The algorithmic pseudocode used in this book is inspired by this language. However, it is not, of course, assumed that the reader knows this language, and we give in this appendix exactly what is necessary to understand the algorithmic expressions used here. It is only assumed that the reader has a minimum of familiarity with programming in general.

We are going to give the basic explanation first using a significant example of an algorithm borrowed from Chapter 5 (backtracking), which expresses the iterative version of the depth-first search algorithm of an arborescence. The explanations given do not require the reader to have previously studied this algorithm. They are simply dealing with the expression of the algorithm.

A.1 Algorithm

```
procedure dfs_arbo_ite(T,r);
begin
  push(S,r);
  v:= r;
  loop
    while exits_child(v) loop
      v:= first_child(v);
      push(S,v);
    end loop;
    -- exits_child(v) false
    while v ≠ r and then not exists_following_sibling(v) loop
      pop(S);
      v:= top_stack(S);
    end loop;
    -- v = r or exists_following_sibling(v) true
    exit when v = r;
    pop(S);
    v:= following_sibling(v);
    push(S,v);
  end loop;
  pop(S);
end dfs_arbo_ite;
```

A.2 Explanations and commentaries

- The words in bold, such as **procedure**, **begin**, **loop**, **while**, **end**, **and**, **then**, **not**, etc., are the *reserved words* of the language, forbidden for any other use than those, generally classic, uses for which they are defined.

- An algorithm will often, as is the case here, be given in the form of a *procedure*, that is a subprogram which has an action on data passed into parameters. However, there are also subprogram *functions* which return a value, introduced by the reserved word **function** instead of **procedure**. When there is only one procedure, as here, it is automatically the main procedure, that is the one by which the running is launched. More generally, a program is made of several subprograms, procedures, and functions, one of them playing the role of the main subprogram.[1]

- The *control structures* used in this algorithm are classic:
  - The *conditional* **if ... then ... else ... end if**. Let us indicate that it is possible, more generally, to add before **else** as many **elsif** as necessary to consider additional cases. Let us note in passing how the structure is bracketed by **if** at the beginning and **end if** at the end. This form is general for all control structures. We find the same method of bracketing for the program itself with **begin** at the beginning and **end** at the end, followed, for ease of reading, by the name of the program.
  - The *loop* **while ... loop ... end loop**, is a well-known structure. However, here we also use a more general loop structure which is better for certain algorithmic constructions: **loop ... end loop** with **exit**. This is the case for the main loop of the procedure. There may be only one exit, as here, but there can also be several, as many as needed depending on the required conditions. Let us note that loop **while** is the particular case of a loop with a single exit under a condition tested at the beginning. Each exit of a generalized loop may be commanded by the statement:

    exit when `(condition);`

    as in the previous algorithm, a form which has the advantage of highlighting the exit condition. However, an exit may also be commanded by the statement **exit** in the conditional **if**.

- In our presentation, the whole environment of the algorithm is supposed to be defined by the context. There are many things here. First, the parameters passed into the procedure `dfs_arbo_ite`, an arborescence `T` and its root `r`. Then, a proper variable of the procedure, `s`, called a *local variable*, here of the type vertex of the graph. Of course, if we wanted to continue to the programming of this algorithm, we would have to have properly declare this variable and its type. Finally, still concerning the environment of the algorithms, we observe the calls, to what can

[1]This is the *main* function in C for example.

be called the *primitives*, that is, subprograms which are assumed to have been given elsewhere and which define the arborescence (here: **`exists_child`**, **`first_child`**, **`exists_following_sibling`**, **`following_sibling`**). Their interpretation is explained in Chapter 5; once again we are interested here solely in the formal algorithmic aspect. We can imagine these primitives gathered in a package (in ADA for example) which the program would call upon. There are other primitives which we will now explain.

- This algorithm uses the classic data structure *stack*. This well-known data structure is handled by subprograms that should be considered as given elsewhere, in a package the use of which will be declared by the program. Since we often use this stack structure (we also use the *queue* structure which can be similarly defined), let us give the formal specifications for these subprograms:
  - `push(S,s)` is a procedure which puts at the top of stack `S` the vertex of the graph contained in variable `s`.
  - `pop(S)` is a procedure which removes from stack `S` the vertex of the graph which is at the top of the stack. In this procedure, and in the previous one, we must consider parameter `S` as being in input–output mode since stack `S` is modified by the action of this procedure (mode **in out** in ADA). Let us specify that this procedure does not return the removed vertex.
  - `top_stack(S)` is a function which returns the vertex of the graph which is at the top of stack `S`. In this subprogram, parameter `S` needs only to be in mode input (**in** in ADA, default mode), since the stack is simply read (in general, a function parameter must always be in mode **in**). In particular, this function does not remove the top of the stack.
  - The Boolean function `is_empty(S)`, unused here, returns *true* or *false* depending on whether or not stack `S` is empty.

We will always additionally suppose that a stack (or a queue) when used has been previously initialized to the empty state.

Let us specify that the data structure stack is general with regard to the elements pushed, here the vertices of the graphs. A *generic* structure may be written in ADA, usable after *instantiation* on the vertex type defined with the type graph.

Let us end these explanations with the following general points:

- Each statement ends with ';', this includes the heading of the program which may considered as a declaration statement of the program and which is called its *specification*.
- Each *comment* is preceded by a double hyphen -- and runs to the end of the current line (that way it is not necessary to close a comment).
- The *assignment* operator := and the *equality* operator =, as others used occasionally, are well known and will not be detailed here.
- It is almost unnecessary to recall the importance of *indentation* in the presentation of a program. It is useful in helping to understand the organization while sustaining visually each control structure.

A.3 Other algorithms

```
procedure dfs_arbo_recu(T,v);
begin
  if exists_child(v) then
    u:= first_child(v);
    dfs_arbo_recu(T,u);   -- recursive call
    while exists_following_sibling(t) loop
      u:= following_sibling(u);
      dfs_arbo_recu(T,u);   -- recursive call
    end loop;
  end if;
end dfs_arbo_recu;
```

A.4 Comments

From the point of view of algorithmic expressions, there is no novelty here compared to the previous example. The novelty lies in the *recursion* apparent here by the calls in the procedure to itself. These calls are signalled by comments. This algorithm is in fact the recursive version of the previous one, called by opposition an *iterative* algorithm. It must be called by a main procedure, as we saw in Chapter 5.

It is not necessary to insist on the importance of recursion in algorithmic expressions and programming! It is a fundamental concept from a practical and theoretical viewpoint (in particular in systems), and this example shows the conciseness of expression allowed by recursion.

Appendix B

Bases of Complexity Theory

B.1 The concept of complexity

In concrete terms, the concept of time complexity corresponds to the time required to run a program. It is an essential practical parameter in numerous applications. This running time depends on many different factors, first of all on the size of the data to be processed. It is obvious that it will not take the same time to process a graph with 100 vertices as one with 1000. The running time has to be considered in relation to the size of the case dealt with. We talk of the *complexity function*.

Another equally essential factor is the power of the computer used for processing. Again, differences may be great, therefore a reference machine has to be specified as we will see. There are other factors, less obvious but as important, such as the manner in which the data are represented. This representation may be more or less efficient with regard to processing.

To speak of complexity means to speak of concepts which at first may seem clear intuitively but which in fact need to be specified and formalized. Let us start with the concept of an algorithm. A complete formalization of this concept requires the definition of a machine in the sense of a model capable of an automatic process. The model which is the principal reference is the *Turing machine*, which bears the name of its inventor in the 1930s. This theoretical "machine" is, as a machine, reduced to its simplest expression, and that is precisely what makes it useful from a theoretical point of view. Also, despite its extreme simplicity, it seems to contain all the calculation possibilities of the most evolved computers. We will not develop this theoretical model here; it is enough to know that it exists and makes it possible to formalize the concept of algorithms.

A machine, no matter which one, processes data which have to be presented to it in a certain manner. For example, graphs may be modeled in different ways. We speak of data *encoding*, which is also a concept which has to be specified because it has a direct influence on the complexity of the processing. Let us consider a simple case with the classic algorithm used to decide if a given integer n is a prime number, meaning that it has no other divisor than 1 and itself. A classic method is to try as divisors all the integers less than or equal to $\sqrt{n}$. The number k of divisions to be done, equal to the integral part of $\sqrt{n}$, is a measure of the complexity of this test since it is clear that the time necessary will be proportional to this number, while considering, nevertheless, the division as an elementary operation of constant time. We may consider *a priori* that this complexity is reasonable since it is simply proportional to the square root of the integer. In fact the value of k is exponential in relation to an usual encoding of integer n. Indeed, in any number system, decimal or binary for example, n is represented by a number of digits proportional to $\log n$ (log being in the base considered), and $\sqrt{n}$ is expressed exponentially in function of $\log n$. In base 10, for example, $\sqrt{n} = 10^{\frac{1}{2}t}$, where $t = \log_{10} n$ is the size of the representation of integer n. To take n as the reference size of the data n corresponds to what is called the unary representation, encoding which consists of using only one digit (a "stick", for example, as in primary school), and in which each integer is written by repeating this digit as often as the value of the number. This manner of proceeding is not adequate in relation to complexity. This encoding is therefore considered "unreasonable". We will suppose implicitly that, in the development which will follow, the encodings used are "reasonable", which is necessary for a realistic complexity concept.

From a general algorithmic perspective, running time is measured by the number of operations which are said to be *elementary*. But this concept of elementary operation depends on the operating level from which we operate, and, above all, on the nature of the problem and the calculation. This can be arithmetic operations, for a sorting algorithm it will involve comparisons and exchanges of elements, and for the processing of graphs visits and specific operations on the vertices. We introduce the general concept of elementary operations *pertinent* for the problem under consideration. These are the operations which are directly involved in seeking the result.

Size of problem	Complexity function		
	n	n^2	2^n
10	0.01 μs	0.1 μs	1.024 μs
20	0.02 μs	0.4 μs	1.049 ms
30	0.03 μs	0.9 μs	1.074 s
40	0.04 μs	1.6 μs	18.3 minutes
50	0.05 μs	2.5 μs	13.0 days
60	0.06 μs	3.6 μs	36.6 years
70	0.07 μs	4.9 μs	374 centuries

Table B.1. *Comparison of complexity functions*

However, we may then worry about the possibly arbitrary nature of this concept, and, even more, about the imprecision which results from neglecting other more elementary operations, or even at the lowest level, "machine operations". In fact, this has no impact on the complexity classes which are defined. Indeed, it is always possible to consider that an operation at a higher level is equivalent to a bounded number of lower level machine operations.

B.2 Class P

After all these specifications, we come to what is at the heart of the complexity theory, that is the *polynomiality* criterion. When the size n of the datum increases, there is a great difference theoretically, but also practically in most cases, between the growth of a complexity function which would be of the order of a polynomial function and an exponential growth. Let us consider for example Table B.1, which gives the running time requested for a datum of size $n = 10, 20, \ldots, 70$, with the number of operations expressed by different complexity functions. Here the hypothesis is of a machine which runs 10^9 operations per second, that is, a billion operations per second, which is already a respectable speed (abbreviations used: s for second, ms for 10^{-3} second and μs for 10^{-6} second; the other time units are spelled out). We see in this table that the running time remains reasonable with polynomial functions n and n^2. However, they no longer remain reasonable with an exponential function such an 2^n as soon as the size n becomes a little large (although still very modest in practice; here we will stop at $n = 70$: the time obtained, 374 centuries, is sufficient explanation!). Asymptotically, it is well known that an exponential function (with a positive base) has a faster growth than any power function.

There is another way to look at things, maybe more explicit. Given a computation time available on a first machine, and an equal time on a second machine 10 times faster, how much bigger is the datum we can process with the second machine relative to the first one? Specifically, let n be the size of the problem which can be processed by the first machine in the given time, and let n' be the size of the problem which can be processed on the second machine for the same available time. Starting first with a complexity function equal to n^2, we have:

$$n'^2 = 10n^2$$

and we deduce:

$$n' = \sqrt{10}n \simeq 3,16n$$

Thus, with the machine which is 10 times faster we can in the same given time process problems which are three times larger in size, which is interesting. As we are going to see, the situation is quite different with an exponential complexity, for example 2^n. We then have:

$$2^{n'} = 10 \times 2^n$$

which gives:

$$n' = n + \log_2 10 \simeq n + 3,3$$

A machine which is 10 times more powerful can only process data of a few additional size units. For an initial datum of size 1000 for example, the gain is not significant.

This polynomial growth criterion of the complexity function was therefore introduced, in particular by J. Edmonds in 1965. This criterion quickly turns out to be pertinent, not just because of its asymptotic nature which we mentioned. There is also the stability of this criterion relative to the composition of algorithms, because of the fact that the composition of polynomial functions is itself a polynomial function. In addition, there is the fact, previously mentioned, that reasonable data encodings only differ from one another polynomially, meaning that any encoding may be upper bounded in size by a polynomial in function of another encoding.

We are used to expressing the complexity of an algorithm with the classic *notation of Landau*: an algorithm is of complexity $O(f(n))$ when the number of elementary operations in relation to the size n of the data is upper bounded by $f(n)$ multiplied by a constant, as soon as the integer n is greater than a certain value. If function f is polynomial, the algorithm is called *polynomial* and the problem dealt with by this algorithm is also called *polynomial*. These problems define the complexity class denoted **P**. Since a polynomial function behaves like its terms of higher degree, and taking into account the constant involved in O, we replace $f(n)$ by an expression of the form $O(n^k)$. Thus we commonly write a complexity in the form $O(n^k)$. The particular case $k = 1$ is in principle the best possible *a priori*, since we need to count at least the time to read the data. This is the case of *linear* algorithms.

Concerning other cases, practice shows that we rarely go over small values of exponent k: $2, 3, 4 \ldots$. This fact reinforces once again the interest of the polynomial criterion: indeed we did mention that an algorithm which was polynomial but with a very high exponent k, for example 10^{50}, would be without practical interest, which is obvious. On the other hand, an algorithm of exponential complexity but with a very low coefficient in front of the exponent, for example $2^{10^{-50}n}$, would not be so bad. We must also say that what is under consideration here is what we call complexity *in the worst case*. This means that we upper bound all data cases uniformly. Yet, it may happen *a priori* that most cases only require a reasonable time, contrary to a few cases which are rare or artificial and are sometimes called "pathological". For such a case, and in general, it seems more natural to evaluate what is called the complexity *in the expected case*, which takes into account appearance rates of each data case, that is the probability with which each possible instance is likely to occur as an input. However, such a measure of complexity is often delicate to calculate, if only because of the fact that it means knowing the input distribution.

A finer analysis of complexity leads us to consider mathematical functions with intermediary growth between the integer-power functions. Thus we often consider function n^q, where q is a real number (not necessarily an integer), $\log^k n$, where k is an integer. (We do not specify the base of this logarithm but we can always consider it to be base 2. It will not change the result expressed asymptotically in O because, as we know, all systems of logarithms are proportional.) To give an example, which does not come from graph algorithms but is sometimes useful, the best sorting algorithms are of complexity $O(n \log n)$, which is better, for example, than $O(n^2)$.

B.3 Class NP

When a problem is recognized as class **P**, it is therefore considered as satisfactorily solved algorithmically. This is the case for numerous basic graph problems such as the search for connected components. However, other problems, apparently simple, at least to set, cannot be solved polynomially.

This is the case with the problem of graph isomorphism. Given two simple graphs G and H, is there an isomorphism of G to H, that is a bijection from vertex set of G onto vertex set of H preserving the neighboring relationship defined by the edge? Let us specify that such a problem with a "yes" or "no" answer is called a *decision problem* and that here we are only considering this type of problem. One way, of course non-polynomial in complexity, is to try the $n!$ possible bijections, n being the common number of vertices of the graphs considered, until finding one which respects the condition. If all bijections have been considered and none is appropriate, then it is possible to answer "no". Let us note an important fact here: given a bijection on the vertices, it is possible to verify polynomially that it does (or does not) define an isomorphism, a complexity $O(n)$ algorithm being easy to imagine for this check. If we have such a bijection, we can say that we can verify polynomially the answer "yes", and the bijection plays the role of a "certificate" for this answer, in the sense that we can be assured of a positive response. If, on the other hand, the answer is negative, then such a certificate does not exist.

We have here, therefore, a problem for which we can verify polynomially, thanks to the certificate, a positive answer, even when this certificate itself cannot be found polynomially. This is the idea which presides over the definition of class **NP**: problems for which it is possible to check a positive answer polynomially without necessarily being able to *find polynomially* this answer. The acronym **NP** does not mean "non-polynomial" but "non-*deterministic* polynomial": the non-determinism here represents our incapacity (which may be temporary) to find directly, that is without the help of a certificate, the right answer. This idea, a bit disconcerting for beginners, is clearly formalized by the non-deterministic Turing machine, but that goes beyond this simple introduction to complexity.

The concept of a certificate, more intuitive than the non-deterministic one, gives a good overview of the concept of class **NP** provided it is formalized. In the following case, we will call an *instance* of a problem a data case for that problem, for example a pair (G, H) of graphs for the isomorphism problem of two graphs. Therefore it is said that a decision problem Π is in class **NP** if there is a polynomial algorithm $\mathcal{A}$ and a polynomial p such that for any instance x of $\mathcal{A}$ the answer is "yes" if and only if there is a datum y such that $|y| \leq p(|x|)$ and algorithm $\mathcal{A}$ applied to x associated with y gives the answer "yes" ($|x|$ designates the size of x, likewise for $|y|$). The algorithm $\mathcal{A}$ is the *checking-algorithm* and y is a *certificate*, for instance x of Π. Let us note the obligation for the certificate to be of polynomial size in relation to the size of the instance; this is an indispensable condition for a polynomial time check. The certificate is said to be *succinct*.

We can immediately verify the inclusion of class **P** into class **NP**: a polynomial algorithm is also a checking algorithm of a class **P** problem, with an empty certificate. On the contrary, one may think that this inclusion is strict, that is that **P** $\neq$ **NP**, taking into account the lower requirement that represents the simple checking of a certificate compared to that of finding a certificate. It is possible, nevertheless, as no one so far has been able to prove or disprove it.[1]

B.4 NP-complete problems

Naturally, from the perspective of the question **P** $\neq$ **NP**, research has been conducted to better understand class **NP**, in particular to spot the most difficult problems of this class in order to attempt to "capture" what creates the intrinsic difficulty of problems of this class.

The comparison tool here is the *polynomial reduction.* A problem π_1 can be polynomially reduced to a problem π_2 if there is a polynomial algorithm, called a *reduction-algorithm* from π_1 to π_2, which calculates for each instance x_1 of π_1, an instance x_2 of π_2 such that the answer for x_1 is "yes" if and only if the answer for x_2 is "yes" (x_1 and x_2 have the same answer). Thus, if we have a polynomial algorithm solving π_2, we can deduce a polynomial algorithm solving π_1, by composing the reduction-algorithm form π_1 to π_2 and the solving-algorithm for π_2 (note the advantage of being able to compose polynomial algorithms). In other words, if π_2 is in class **P**, so is π_1. Again, π_2 is at least as difficult as π_1 and is thus possibly more representative of the difficulty of class **NP**. The final interest of all of this is to put in evidence a problem at least as difficult as all the others in class **NP**, that is

[1]There is a one million dollar reward offered by an American patron for whoever solves the problem. Go for it!

a problem to which all others can be reduced polynomially. A "universal" problem, in a way, for class **NP** is called **NP**-*complete.* Such a problem can be formally put in evidence with the non-deterministic Turing machine mentioned above. However, it is more interesting to know that there are such problems which can be put naturally. The first found, at the beginning of the 1970s, is a problem of logic, called a *satisfiability problem* (denoted SAT) which we will now describe.

The data are: n Boolean variables $x_1, \dots, x_n$ taking the value *true* or *false*, m Boolean expressions $C_1, \dots, C_m$ called *clauses* and expressed in a disjunctive form, that is:

$$C_j = y_1 \vee \cdots \vee y_{k_j}$$

where each y_i is equal to x_l or $\neg x_l$, where $l \in \{1, \dots, n\}$ (remember the classic logic operators: *or* denoted $\vee$, *no* denoted $\neg$). The question then is: is their an assignment of values to variable x_i such that each clause C_j takes the value *true* (that is it has at least one y_i which has the value *true*)? Let us note that this problem is clearly in class **NP**: such an assignment is in fact an appropriate certificate. Cook–Levin's theorem states that this problem is **NP**-complete. Historically, SAT has been the first natural problem found in class **NP**, but there have been numerous others found since, in particular concerning graphs.

B.5 Classification of problems

It is impossible to speak of the bases of complexity theory without mentioning the class **coNP** and the concept of a "well-characterized" property. The definition of this class is based on the dissymmetry which exists between the answers "yes" and "no" in the problems of class **NP**. This dissymmetry does not appear in class **P**: the answer "no" is automatic if it is not the answer "yes". For a problem of class **NP**, the answer "no" may not have an obvious succinct certificate. For example, for the problem of graph isomorphism, what can be a certificate for the answer "no"? Another example: for the problem of the existence of a Hamilton cycle in a graph, if it is easy to see how to "certify" the answer "yes", simply by giving a Hamilton cycle, we do not see directly how to certify the answer "no".

Class **coNP** is therefore defined as the class of decision problems for which the "complement" problem, that is the problem set in order to invert the answers "yes" and "no", is in **NP**. We do not know if **NP** = **coNP** and, in the hypothesis of an inequality, the intersection of these two classes is interesting to consider: it represents the problems based on a *well-characterized* property in the sense that the answer "yes" as well as the answer "no" can be certified by a succinct certificate. This is a typical case, for example, with the problem of recognition of planar graphs: the answer "yes" can be certified by a planar embedding of the graph, while the answer "no" can be certified by the presence in the graph of an excluded configuration (by application of Kuratowski's theorem, see Chapter 11). However, the problem of the recognition of planar graphs is in fact in **P**; it is therefore not surprising that it belongs to this intersection since we have in general $\mathbf{P} \subseteq \mathbf{NP} \cap \mathbf{coNP}$. Another basic question of complexity theory is to know if we have equality in this inclusion. Very few problems are known in this intersection without also being known in **P**.

On the whole, we have a possible configuration of the previous classes as shown in the diagram in Figure B.1. Some people think this is probable. If it is not like this, then things are very different from what they are thought to be today ....

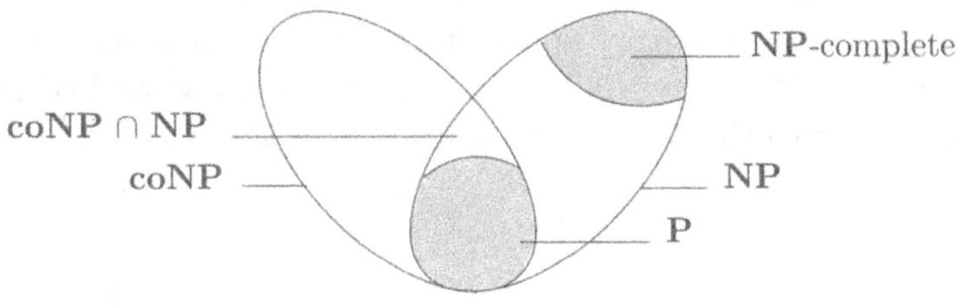

Figure B.1. *Classes of complexity*

Let us specify, finally, that an *optimization* problem is different from a *decision* problem. Let us take for example the traveling salesman problem (Chapter 10), which can be stated as follows: find in a weighted graph a Hamilton cycle minimum for the sum of the values of its edges. We associate it with the following decision problem: given an integer k, is there a Hamilton cycle of length $\leq k$? Obviously a solution to the optimization problem gives a solution to the decision problem, simply by comparing the value of a minimum cycle with k. However, what is more interesting is that the converse is true. It is less obvious; it can be seen by bounding by successive dichotomies the value of a minimum cycle. A problem for which the associated decision problem is **NP**-complete is sometimes called **NP**-*difficult* rather than **NP**-complete.

We again find in an interesting way the idea of a well-characterized property in certain optimization problems, typically the maximum flow problem in a transportation network and that of the minimum cut (Chapter 8). At the optimum, we have an equality which certifies one as maximum and the other as minimum.

B.6 Other approaches to difficult problems

Once it has been admitted that some problems can probably not be practically solved in a reasonable manner, that is by a polynomial algorithm (in the hypothesis **P** $\neq$ **NP**), other approaches to **NP**-complete problems have to be contemplated.

A first, natural, approach is to try to obtain in polynomial time an approximate solution with a precision which has to be given. Problems are not all equal when it comes to this. For some it is possible to find good approximations, others are resistant to any reasonable results, such as, for example, the traveling salesman problem for which one shows that, except if **P** = **NP**, there is no satisfactory approximate polynomial algorithm to solve it (see Chapter 10). There are many more approaches, not studied here, and we refer the reader to the specialized literature on this subject.[2]

[2]For example: C.H. Papadimitriou, *Computational Complexity*, Addison-Wesley (1994); J. Stern, *Fondements mathématiques de l'informatique*, McGraw-Hill (1990); M.R. Garey and D.S. Johnson, *Computers and Intractability*, Freeman (1979), a great classic reference in the field.

Bibliography

[1] Berge C., *Graphs and Hypergraphs*, North-Holland Mathematical Library, vol. 6, North-Holland, 1973.

[2] Berge C., *Graphs*, North-Holland Mathematical Library, vol. 6, North-Holland, 1985 (Second revised edition of part 1 of the 1973 English version).

[3] Bang-Jensen J. and Gutin G., *Digraphs Theory, Algorithms and Applications*, Springer, 2001.

[4] Bondy J.A. and Murty U.S.R., *Graph Theory*, Graduate Texts in Mathematics, vol. 244, Springer, 2008.

[5] Diestel R., *Graph Theory*, 3rd ed., Graduate Texts in Mathematics, vol. 173, Springer, 2005.

[6] Gondran M. and Minoux M., *Graphes et algorithmes*, Collection de la Direction des Études et Recherches d'Électricité de France, vol. 37, Eyrolles, Paris, 1979.

[7] Graham R.L., Grötschel M. and Lovász L. (Eds.), *Handbook of Combinatorics*, vol. 1, 2, Elsevier, 1995.

[8] Jungnickel D., *Graphs, Networks and Algorithms*, Algorithms and Computation in Mathematics, vol. 5, Springer, 1999.

[9] Lovász L., *Combinatorial Problems and Exercises*, 2nd ed., North-Holland, 1993.

[10] Lovász L. and Plummer M.D., *Matching Theory*, Annals of Discrete Mathematics, vol. 29, North-Holland, 1986.

[11] Mohar B. and Thomassen C., *Graphs on Surfaces*, Johns Hopkins Studies in the Mathematical Sciences, The Johns Hopkins University Press, 2001.

[12] Prins C., *Algorithmes de graphes*, 2nd ed., Eyrolles, Paris, 2003.

[13] Van Leeuwen J., "Graph Algorithms", Chapter 10 in *Handbook of Theoretical Computer Science, vol. A, Algorithms and Complexity*, edited by J. van Leeuwen, Elsevier, 1990.

Index

CPSIA information can be obtained
at www.ICGtesting.com
Printed in the USA
LVHW022232240721
693592LV00002BA/26

9 781848 210707